Dic 99.

Ivan:

Se que en el
proximo 2000
va a ver un gran
Director de Cine que
haran que Spielberg, Coppola
y J. Lucas se queden en la
historia. Y ese gran director
eres tu! Porque tienes el
Talento y la Inteligencia
necesarias. Solo nunca dejes
de soñar en Grande y trabaja
duro! Hay que levantarse
Temprano para hacer los
sueños realidad! Mis mejores
deseos para el proximo
milenio!
con mucho cariño

LEARN TO MAKE VIDEOS IN A WEEKEND

LEARN TO MAKE VIDEOS IN A WEEKEND

ROLAND LEWIS

Photography by John Bulmer

ALFRED A. KNOPF
New York
1993

A DORLING KINDERSLEY BOOK

This edition is a Borzoi Book published in 1993 by Alfred A. Knopf, Inc.,
by arrangement with Dorling Kindersley.

Art Editor Alison Donovan
Editor Deborah Opoczynska
Series Art Editor Amanda Lunn
Series Editor Jo Weeks
Managing Art Editor Tina Vaughan
Managing Editor Sean Moore
Production Controller Helen Creeke

Library of Congress Cataloging-in-Publication Data

Lewis, Roland
    Learn to Make Videos in a Weekend / by Roland Lewis. -- 1st ed.
        p. cm. -- (Learn in a weekend series)
    Includes index.
    ISBN 0-679-42230-7 :
        1. Camcorders--Amateurs' manuals. 2. Video recordings--
Production and direction--Amateurs' manuals. I Title.
    TR882.L49 1993
        778.59'92--dc20   92-54780
                                                             CIP

Computer page make-up by Cloud 9
Designs, Hampshire, and The
Cooling Brown Partnership,
Hampton-upon-Thames
Reproduced by Colourscan,
Singapore. Printed and
bound by Arnoldo
Mondadori, Verona, Italy

First American Edition

CONTENTS

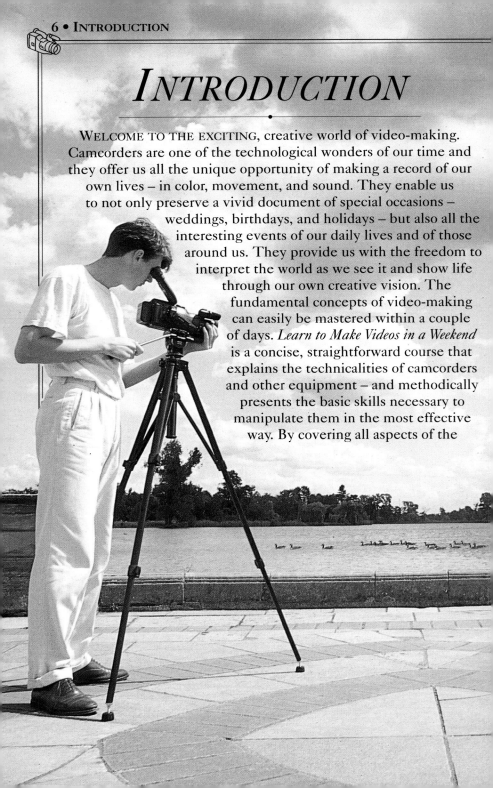

INTRODUCTION

WELCOME TO THE EXCITING, creative world of video-making. Camcorders are one of the technological wonders of our time and they offer us all the unique opportunity of making a record of our own lives – in color, movement, and sound. They enable us to not only preserve a vivid document of special occasions – weddings, birthdays, and holidays – but also all the interesting events of our daily lives and of those around us. They provide us with the freedom to interpret the world as we see it and show life through our own creative vision. The fundamental concepts of video-making can easily be mastered within a couple of days. *Learn to Make Videos in a Weekend* is a concise, straightforward course that explains the technicalities of camcorders and other equipment – and methodically presents the basic skills necessary to manipulate them in the most effective way. By covering all aspects of the subject, it is a comprehensive guide to the abilities you can develop to make effective, entertaining, and memorable videos. It provides both a solid grounding in the basic skills of video-making and an initiation into the imaginative and creative techniques that can make your videos distinctive. When you first use your camcorder, shoot exploratory exercises using the weekend course as a guide. This will enable you to test your own aptitude. Show the results to friends and other camcorder users to check their reactions and keep referring back to the book to refine your skills still further. Experiment with the creative techniques illustrated so that they become a natural part of your video-making repertoire. Above all, get into the habit of using a camcorder as much as possible. Not only will the practice give you the pleasure of proficiency but the custom of regular shooting will provide challenges for your creative imagination – and deep satisfaction with the experience.

Roland Lewis

ROLAND LEWIS

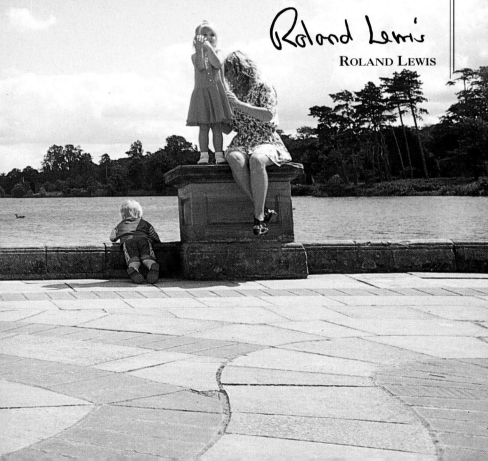

PREPARING FOR THE WEEKEND

Choosing and getting to know the capabilities of your equipment

•

THE SIMPLEST AND CHEAPEST, "point and shoot" camcorders tend to be fully automated and are suitable for everyday family events. For greater creative potential, look at models that will give you the additional option of manual control of key functions. However, if you intend to make video-making your major hobby or want to produce more professional work, then investigate the top end range of equipment with the maximum of controls and facilities. Take time over your choice so that you get a model which suits your intended needs. Go through the functions explained in your user's manual carefully *before* the weekend. Carry out plenty of

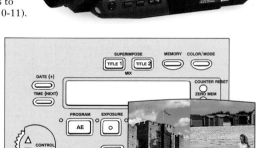

CAMCORDER DESIGN

Camcorders are available in different types of format, according to the different tape size, cassette design, and recording system. They also vary in design from lightweight palmcorders, through compact models to more advanced, full-size models (pp.10-11).

CREATIVE TOOL

Camcorders concentrate complex optics and electronic functions into one compact unit, that can record images and sound, but also allows for personal creativity (see pp.12-13).

FUNCTIONS

Many camcorders bristle with control buttons and special facilities to help embellish your recordings. The most advanced of these models incorporate digital processing to produce sophisticated, more elaborate, visual results (see pp.14-15).

test shooting to develop your basic skills before launching into the first important subject you may want to record. Video tapes are comparatively cheap, but more importantly, you can re-record over unwanted material so test shooting will cost you nothing yet will pay dividends as soon as you start shooting in earnest.

Once you have had sufficient experience using your camcorder, consider which accessories will best serve your needs.

*Words in **bold** are explained in the glossary.*

ACCESSORIES

The kind of accessories you need depends on the topics you wish to shoot, but overall, the most useful equipment for all video-makers will be a tripod, lens converters, lights, and additional mikes (see pp.16-17).

EXPOSURE

All camcorders have auto-exposure systems that adjust the aperture to suit lighting conditions. Many models are also fitted with a "gain" facility that is able to boost the image by producing an apparent increase in the **exposure** (see pp.18-19).

COLOR & CONTRAST

You can achieve perfectly lighted and colored pictures in all conditions so long as you are aware of the different factors that determine the way in which color is reproduced (see pp.20-21).

AUTO-FOCUS

Auto-focus is a feature common to most camcorders. Although helpful in many situations, the systems can be fooled into producing the wrong results – for instance, where the main subject is not at the center of the frame, or when shooting very fast moving subjects. Manual controls can help solve most of these problems (see pp.20-11).

WHICH CAMCORDER?

Evaluating the main differences between the various types of camcorder

ALL CAMCORDERS SHARE the same basic functions. They can record moving pictures in color with sound. They are equipped with a viewfinder that acts as a monitor, enabling the immediate playback of the material which has been recorded. They can be powered by re-chargeable batteries, from an electrical outlet, or from a car battery. All models are equipped with a variable **focal length (zoom)** lens and a built-in microphone. The main variations between the different machines relate to the format (tape size), band (image resolution and quality), and the sound system, as well as the design, shape, and weight. The subsidiary differences have to do with the degree of automation of functions and the special facilities the camcorders possess (see pp.14-15).

FORMATS

The most common format camcorders use is 8mm video. For replay, the camcorder must be connected to the TV, or alternatively, an 8mm VCR can be used. VHS-C is a compact version of the format used for home VCRs. Replay involves slipping the cassette into a VHS adaptor. VHS is the original camcorder system and is compatible with most domestic VCRs. This cassette size dictates the larger, usually heavier types of machines.

Adjustable viewfinder •

• Built-in mike

PALMCORDERS
Palmcorders are the smallest and lightest of the camcorders and use 8mm or VHS-C cassettes. The simplest "point-and-shoot" versions rely on automated functions. The more expensive machines incorporate sophisticated additional facilities and allow for greater manual control of key functions.

COMPACT
The mid-size compact camcorders also use either 8mm or VHS-C cassettes and range from simple versions with limited functions to models with a variety of additional facilities. The body is sturdier but heavier than that of the palmcorder's. The greater surface area allows for larger control buttons.

FULL-SIZE

The full-size machines are likely to possess the greatest range of facilities and offer the most control over basic functions. They will be larger and more expensive, whichever format they use. Due to the extra weight of these models, they are designed to sit comfortably on the shoulder while you are shooting.

Carrying handle •

Motorized • zoom lens

HIGH BAND

All three formats are available as high band versions (S-VHS, S-VHS-C, and Hi8). With the high band format, the luminance and chroma signals are recorded separately to reduce any color interference and improve the picture quality by increasing the image resolution to about 400 lines – the standard is 250 lines. The full improvement, however, is only possible when the tapes are played back through a monitor with an **S-Terminal**. Because this recording system is different from the usual standard version, tapes can only be played back on high band VCRs.

TAPES
The camcorder cassette sizes and maximum playing time of the tapes vary between the three different types of format.

• S-VHS tape

Hi8 tape •

• Adaptor

ADAPTOR
S-VHS-C and VHS-C tapes use a simple adaptor for playing back tapes using a VHS-type VCR. The lid is lifted and the cassette is slotted into the niche.

• S-VHS-C tape

CHOOSING A CAMCORDER

Apart from the price, there are a number of other important factors to consider:
• Weight and portability – particularly if you plan to do a lot of travel videos.
• Picture quality – important when you plan to copy tapes to give to other people.
• Tape length – only significant if you have subjects that demand it. A half-hour tape is quite adequate for most subjects, whatever the format.

• Robustness and durability – the larger, full-size machines tend to be better suited to prolonged, heavy use.
• Convenience – compatibility with your home VCR system.
• Sound quality – mono and stereo models are available. The 8mm models offer high quality **FM** recording, but generally, the sound track cannot be replaced except as part of an editing or copying process.

HOW IT WORKS

Understanding the inside workings of the optical and recording processes

A CAMCORDER CONSISTS of two main elements – a camera and a video recorder – both contained within a single compact unit. These machines incorporate complex optical devices, sophisticated electronic processes, and ingenious recording mechanisms. The degree of miniaturization of the components means that they are constructed as precision instruments. Although knowledge of the intimate operations of the camcorder is not essential, an understanding of the key processes can be useful in exploiting their creative potential.

CAMERA SIDE

Basic camcorder functions are usually fully automated, although on some models there will be manual overrides. Auto-iris controls the amount of light that reaches the image sensor, ensuring correct **exposure**. Auto-focus uses light or infra-red beams to adjust the focus as you shoot. Auto white balance (AWB) is used to ensure correct color reproduction. Auto level control (**ALC**) sets the sound recording level while you shoot, and adjusts volume according to the source.

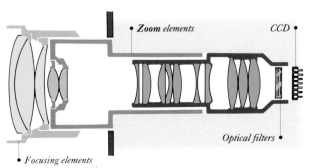

• *Zoom elements*

CCD •

• *Focusing elements*

Optical filters •

LENS

The image from the lens is focused (via a set of optical filters) onto the charge coupled device (CCD) image sensor. This sensor is made up of a matrix consisting of up to almost half a million picture elements called pixels. Pixels analyze the image in terms of color and brightness, and convert this information into an electrical current.

VIEWFINDER

During recording and playback, signals are sent to the miniature cathode ray tube in the viewfinder. In some models, the viewfinder has a color LCD (liquid crystal display).

RECORDING

When the power is turned on, the camcorder is in the record/pause mode with the head drum spinning. As the record button is pressed, the tape advances and within a second or two, the recording starts, continuing until the record button is pressed again. A **flying erase head**, also set on the drum, precisely erases the tail of each recording before the new one starts. This ensures that there is a disturbance free "cut" from one shot to the next. The small loss in the erased recording must be taken into account when shooting (see p.59). Each video frame consists of signals from the two heads. A **synchronizing pulse** is recorded as a separate track on the tape at the same time and governs the playback speed, **helical scan**, and head drum synchronization.

• *Video cassette*

• *Magnetic tape*

• *Capstan*

Playback • • *Weight*
head

• *Pinch rollers*

Sound • • *Record head*
head

• *Flying erase head*

HEAD DRUM

On the recorder side, the slow moving tape is wrapped around one half of a rapidly spinning drum that contains the two recording heads. As the tape passes, the heads convert the amplified electronic current to record the pictures in the form of magnetic patterns. The drum is set at an angle, so that the image is "written" diagonally across the tape.

SOUND

Sound is either recorded with a fixed head, as a linear strip along one edge of the tape, or by another head on the drum, which records the sound enmeshed with the pictures.

PLAYBACK

Since the viewfinder acts as a monitor, the recording can be played back and checked through it at any time. By using headphones, you can also check the sound quality.

CAMERA CONTROLS

Learning the functions and uses of the various camcorder facilities

APART FROM THE BASIC FUNCTIONS, such as the **zoom** and auto-focus facilities, all camcorders offer a range of additional built-in facilities and are able to accept specific accessories. It is improbable that any single model will possess the whole range of features that are currently available, but many useful facilities are likely to be found on even the most modest camcorder. Although these special features are not essential, they can often embellish the visual style of your videos.

Auto-focus •

— CAMERA FACILITIES —

A range of the useful in-camera facilities that are available in many current camcorders.

TITLE SYSTEM
A built-in title system which generates both letters and colors, and also enables scrolling.

TIMER
Interval timer allows very short bursts of recording to produce **time-lapse** effects.

MANUAL ZOOM •
The manual **zoom** lever enables you to override the zoom motor, providing you with greater control over your zoom movements – particularly useful when shooting action subjects.

ANIMATION
Set the camcorder to record just a few frames at a time to produce **animation**.

EFFECTS
Digital processing is the most advanced feature for producing special visual effects.

Start in close-up to emphasize the girl's expression

Zoom out to reveal reason for the girl's happy expression

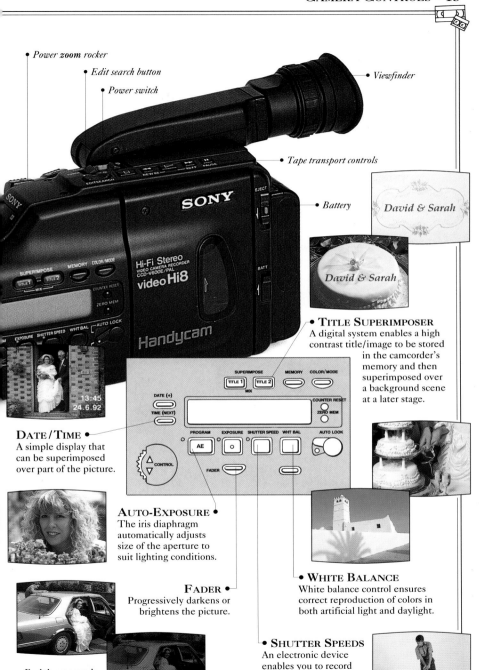

- *Power **zoom** rocker*
- *Edit search button*
- *Power switch*
- *Viewfinder*
- *Tape transport controls*
- *Battery*

David & Sarah

David & Sarah

• TITLE SUPERIMPOSER
A digital system enables a high contrast title/image to be stored in the camcorder's memory and then superimposed over a background scene at a later stage.

DATE / TIME •
A simple display that can be superimposed over part of the picture.

AUTO-EXPOSURE •
The iris diaphragm automatically adjusts size of the aperture to suit lighting conditions.

FADER •
Progressively darkens or brightens the picture.

• WHITE BALANCE
White balance control ensures correct reproduction of colors in both artificial light and daylight.

End the sequence by fading out the shot till screen is black

• SHUTTER SPEEDS
An electronic device enables you to record images of a short period – as little as one ten thousandth of a second.

EQUIPMENT

Expected use will determine your choice of extra equipment

A BASIC CAMCORDER is all you need to shoot successful videos. However, there is a range of accessories that can help improve the technical quality of your recordings, extend your creative scope, or simply protect your equipment. Camera stability is a high priority since it will effect all your shooting. Tripods are the ideal solution although weight and convenience can make monopods, chest, and shoulder pods worthy substitutes. Sound is also important. As built-in mikes have limited capabilities, most camcorders have both an accessory shoe and mike socket for supplementary mikes. Headphones, filters, bags, lights, and converter lenses should also be regarded as natural video accessories.

Palmcorder bag

Aluminum case

Compact shoulder bag

Tripod

BAGS

Soft, lightweight, shoulder bags are ideal for palmcorders. Make sure there is space for a spare tape and battery. A sturdy hold-all with accessory pockets will accommodate compact camcorders of any design and provide ready access to your equipment. Choose one with waterproof material. Aluminum cases with padding inside provide the best protection.

Chest pod

Fluid head

TRIPODS

Tripods offer essential support enabling you to take reliably steady pictures. Choose models with braced legs and always opt for sturdiness over lightness. A **fluid head** ensures smoother camera moves and a spirit level aids level shots. Monopods are not so stable but save on weight.

Shoulder brace

Monopod

• Light mounted on camcorder

Mains light •

• On-board battery light

• Light with separate battery pack

Battery light •

BATTERY LIGHTS
Battery lights are not just for illuminating low light scenes, they can also help improve the picture quality in many indoor situations. Generally, the more powerful the light, the more it will weigh. These lights are most effective for taking close-ups and for small groups, such as at an informal party.

Teleconverters •

Wide angle • converters

CONVERTER LENSES
Converter lenses are optical attachments that increase the telephoto or **wide angle** range of your **zoom** lens. Wide angle converters are particularly useful as they compensate for the limited angle of view provided by almost all zoom lenses. Check their effect on your auto-focus and motor before buying.

• Graduated filters

FILTERS
A skylight or ultraviolet (UV) filter will eliminate the blue haze produced on outdoor shots. It will also provide a good form of protection for the front element of your **zoom** lens and it can be fixed into place.

• Polarizing filter

Ultraviolet filter •

MICROPHONES
Try extending the range and quality of your recordings by choosing from the variety of supplementary mikes that are able to override the built-in mike. Fix battery powered mikes to the accessory shoe or use other mikes that can be hand-held or clipped to clothing for interviews.

HEADPHONES
Headphones are an essential requirement to confirm what you are recording on the sound track. Personal earphones will suffice but the larger, closed-back type that cover the ears are better for eliminating unwanted sounds.

Hand-held • mikes

Camcorder • mountable mikes

• Closed-back headphones

• Mike attachment

Tie-clip mikes

EXPOSURE

Ensuring that you record images at the correct level of brightness

ALL CAMCORDERS ARE EQUIPPED with an automated **exposure** system in the form of an iris diaphragm, positioned within the lens system. The iris is under direct electronic control – adjusting the aperture to the shooting conditions. This ensures that the image you record is as faithful a reproduction of the scene as video recording will allow. Auto-exposure systems are a great benefit in video-making since they allow you to concentrate on the other aspects of shooting. There are some situations where the system will misread the lighting conditions. Be aware of the need to compensate to ensure appropriate exposure.

APERTURE

Most of the time, the auto-exposure will adjust successfully to ensure an appropriate aperture setting for the scene you are recording. However, momentary changes in **exposure** will be evident if the video-maker moves the camera quickly across from a brightly lit subject to one that is in shadow, or if the content of your shot changes suddenly.

LIGHTING
In bright sunlight, the aperture closes down to its minimum size – in poor light it opens up to its maximum.

Overlapping leaves • form the aperture

Aperture determines • the amount of light reaching the sensor

IRIS DIAPHRAGM
The iris consists of a system of overlapping leaves that form the aperture. It determines the amount of light reaching the sensor. The system is under direct electronic control.

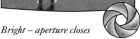

Bright – aperture closes   *Dim – aperture opens*

MANUAL IRIS

Camcorders with a manual iris give the greatest control over **exposure**. Using the viewfinder for reference, you are able to adjust the aperture for exposure for each individual shot. Manual irises may be controlled using a knurled ring or a pair of buttons found on the camcorder body. The different apertures are displayed in the viewfinder screen. This provides you with creative control (for instance, you can deliberately **underexpose** shots for effect) but requires a more concentrated effort. The size of the iris aperture is measured in **f-stops** and these calibrations may be shown on the lens barrel, although usually only the maximum aperture is displayed.

AUTO-EXPOSURE

It is possible to overcome some auto-exposure problems by adjusting the composition. Where the main subject is in a shadow, avoid including too much surrounding, sunlit areas or the subject will be **underexposed** – move in for a closer shot. With shots where the main subject is seen against a large area of dark background, the subjects will appear **overexposed** – change your angle or move closer.

The castle is underexposed

Tilting down corrects the exposure

COMPOSITION
Above: Too much sky makes the castle appear dark. Right: Correct the **exposure** by tilting down to reduce sky.

Dim candle lit shot – underexposed

Bright daylight shot – correctly exposed

Excessively bright shot – overexposed

LUX

Lux is the unit for measuring the amount of light illuminating a given area. Camcorders are able to operate in low light – often as little as three lux or less. This means that you can take shots by candlelight. However, the best results are obtained in much brighter situations such as full daylight. Under these conditions, images will be sharp and colors will be fully saturated and more natural looking.

TOO BRIGHT
In excessively bright conditions, fit a neutral density (ND) filter onto the lens. This will ensure the picture is correctly **exposed**. Use fast **shutter speeds** for the same effect.

BOOSTING EXPOSURE

BLC
When the light is coming from behind the subject, auto-exposure is a problem. The iris closes down and the subject appears in silhouette – such as an interior shot of a person against a window. The backlight compensator (BLC) opens up the iris to correct the **exposure** on the subject.

GAIN
Some camcorders are also equipped with a "gain" facility that electronically amplifies the signal when taking shots in low light situations. Increasing the gain boosts the image to give an apparent increase in exposure. However, this will result in poor image quality – the picture becomes "grainy", less sharp, and the colors become dull and murky.

*Subject **underexposed***   *BLC lights the subject*

CONTROLLING COLOR

Understanding how the camcorder sees light and color

"WHITE" LIGHT IS MADE UP of all the colors of the spectrum and different sources of light produce light of different **color temperatures**. The lowest temperatures are at the red end of the spectrum (candlelight) and the highest are at the blue end (bright, clear days). Artificial sources, including domestic tungsten lighting, produce light of specific temperatures. Our eyes compensate for these differences, but the electronics of camcorders must be set to ensure that white is recorded as white.

WHITE BALANCE

Most current camcorders are equipped with an automatic white balance (AWB) facility. The AWB registers the **color temperature** of the prevailing source and makes the necessary adjustments to ensure correct reproduction of all colors. Some camcorders also have pre-set white balance positions to cope with the different lighting conditions. However, with this system it is crucial that you choose the right setting – if you fail to do so, the picture will have an unnatural color cast.

Tungsten setting results in a bluish cast

Daylight setting records correctly

DAYLIGHT
The daylight setting is used for exterior shots. Daylight scenes shot on the tungsten setting will appear blue-tinged.

Daylight setting results in an orange cast

Tungsten setting records correctly

TUNGSTEN
The tungsten setting is used for lit interiors or shooting with video lights. Lit interiors shot on a daylight setting will have an orange cast.

MIXED LIGHT

Some camcorders can help deal with the problem of mixed light sources by being locked into a position that registers the varied sources. The way this works is by covering the lens with a white cap or pointing it at a white card, so allowing the electronics to lock onto this as a reference for white.

Cover the lens with a white cap or point it at white card

USING FILTERS

Basic color rendering is determined by the electronics of the auto white balance facility, but you can easily change and improve the image with the use of lens filters – providing that you are able to set or lock the AWB on your camcorder. The skylight filter, also known as the ultraviolet filter, is the most valuable filter for video – it provides an effective form of protection for your lens and helps to reduce the blue haze commonly encountered on sunny days.

No filter

Filter used

POLARIZING FILTERS
Polarizing filters increase image sharpness and color saturation. They can also suppress reflections from non-metallic surfaces.

*Graduated (neutral density) filter, balances **exposure** of sky*

NEUTRAL DENSITY FILTERS
Neutral density filters are used to reduce contrast ratios, depth of field, and also the amount of light reaching the image sensor.

Pink filter used to add warmth and improve flesh tones

COLOR FILTERS
Single color filters can enhance scenes by providing an overall color cast. Strong colors can be used for dramatic effects.

Shadow area "fills in" under the umbrella

Diffused even light reduces the contrast

COMPOSITION
If conditions are too contrasting, adjust the composition, take separate shots of the area in shadow, or simply wait for the sun to go in.

CONTRAST

Even when the **exposure** is correctly adjusted, the camcorder may still find it difficult to resolve extremes of light and shade. With shots taken in strong, direct sunlight, the shadow areas that are visible to the human eye tend to "fill-in" – losing the details within them. Sunlight will always add sparkle to your pictures, but duller days with low contrast, diffused light can also produce pleasing results on video. In interiors, you can overcome contrast problems by lighting shadow areas.

FOCUSING

Making sure you always have a sharply defined image

ALL CURRENT CAMCORDERS are equipped with auto-focus systems to help ensure your subjects are sharply focused. These sophisticated, electronic systems control the motorized movement of a focusing element within the lens. One system uses infra-red beams fired from a window beside the lens to measure your distance from the subject. The other, through-the-lens (TTL) method uses an image sensor to detect image sharpness. However, both systems have drawbacks for which you will need to compensate.

AUTO-FOCUS

Most of the time, the systems can interpret what the lens sees and keep the shot in focus, even when the subject is moving, you are moving with the camera, or you are **zooming** the lens in or out. Nevertheless, there will be situations where the auto-focus will be misled and focusing becomes inaccurate. Using the manual focus control will help you to compensate for most of the trickier situations.

FAST SUBJECTS
Fast moving subjects may cause a temporary loss of focus as the system searches for the correct point. A **wider angle** lens shot will help reduce this.

COMPOSITION
Subjects placed at the edge of frame may be out of focus. Use manual focus to lock onto your chosen subject and then recompose.

NETTING
Use manual focus or position the lens through the bars to ensure that the lens focuses on the subject and not on the foreground.

FOCUS RULES

When using manual focus, always check the focus first. Before setting up a shot, **zoom** into the subject, focus the lens then zoom out and compose the shot. Use the telephoto to ensure focusing accuracy. You can also use manual focus for creative effect by starting a shot with one subject in focus and then guiding the viewer's interest by throwing the focus onto a second subject in the frame. Keep your subject clearly in focus whenever you shoot a moving shot. As you move, so will the focus distance, so adjustments will need to be made.

LOW LIGHT
Auto-focus is less effective in low light situations and infra-red systems have difficulty with non-reflective subjects such as the dark hair of a cat.

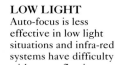

MANUAL FOCUSING

Some camcorders enable you to deal with auto-focus problems by manual focusing of the lens. This does require a certain degree of skill. Your subject is viewed through the viewfinder and the focus ring on the lens barrel is rotated until the image grows sharp. You must learn which way to rotate the focus ring to "follow focus" with a moving subject, particularly when you are using the telephoto position.

Start off shot deliberately out of focus

Subject is pulled into focus

PULL FOCUS
For creative effect, begin with the shot out of focus, then pull in focus to bring the subject into sharp relief.

The narrow depth of field requires accurate focusing

MACRO FOCUS
Auto-focusing is not possible in the **macro** position. Move the camcorder in relation to the subject or use the **zoom** lever to focus.

CLOSE FOCUS

On any lens, there must always be a minimum distance between the lens and the subject. On most camcorders this distance is around one meter (3ft). Anything within this distance will be out of focus. However, the majority of camcorders today incorporate a **macro** position that will enable you to work at very close distances to the subject.

DEPTH OF FIELD

Depth of field is the distance between the nearest and furthest points that will be in focus in any shot. Since controlling the depth of field is an important creative skill, it is essential that you understand the principles that govern it: • The greater the **focal length**, the narrower the depth. • The greater the aperture, the narrower the depth. • The nearer the subject is to the lens, the narrower the depth. • Using fast **shutter speeds** will increase the aperture and reduce the depth. Therefore, the greatest depth of field is achieved with something like a landscape view, shot on a bright, clear day, and with the lens set at its widest setting.

SEPARATION
Telephoto, wide apertures in low light, or fast **shutter speeds** reduce the depth of field and separate subjects from surroundings.

DEEP FOCUS
Using **wide angles**, small apertures in bright light, or slower than normal **shutter speeds** will create a greater depth of field.

THE WEEKEND COURSE

Understanding the weekend course at a glance

•

THE COURSE COVERS fourteen skills in two days. The first day covers skills one to nine. Skills one to three are about getting to know your camcorder – how to hold it securely, and how to move with it. The following six skills cover the basic techniques you need to become familiar with to produce good videos, such as using **zoom** movements and which lights to use for each different situation. On the second day, you will move onto the more advanced, post-production techniques that will enable you to produce "polished" videos you can be proud of.

Using a tripod enables smooth tilts (p.32)

Craning to follow rising or falling movements (p.35)

Simple mike test (p.41)

KEY TO SYMBOLS

CLOCKS
A small clock marks the start of each new skill. The blue section is a guide to how much time you may need to spend learning that skill and the grey segment shows where the skill will fit into your day. The clock does not indicate the time it will take you to apply any particular method, but how much time you might spend practicing each skill until you feel confident.

RATING SYSTEM •••••
The complexity of a skill is rated by a bullet system. Each skill in this book is rated in terms of its relative difficulty on a scale of one to five. The bullets that appear against each skill show how much of a challenge it is likely to be. One bullet (•) denotes a skill that is relatively easy to acquire. A five bullet (•••••) skill is the most challenging and you may not be able to master it within just a single weekend.

Zoom in to emphasize subject's expression (p.37)

Try out different angles for a more interesting composition (p.49)

DAY 2		Hours	Page
SKILL 10	Editing	3	58
SKILL 11	Rostrum shooting	1	62
SKILL 12	Animation	1½	64
SKILL 13	Creating titles	1	66
SKILL 14	Sound tracks	2	70

Raise the camcorder above your head for a high angle (p.49)

Use an accessory mike for recording commentary (p.71)

TÜRKIYE

Adjust the letter coloring to ensure it contrasts sufficiently with its background (p.66)

1

HOLDING THE CAMERA

Definition: *Avoiding camera shake and obtaining stable shots*

THE DELIGHT OF CAMCORDERS is that they are lightweight and portable. Most of the time you will want to take advantage of this feature and shoot hand-held. Camcorder design is geared to encourage this way of operating. All camcorders have a right grip-strap for secure holding. The start/stop trigger is positioned so that it can be operated with the thumb, and the **zoom** "rocker" control is within easy reach of the first and second fingers. The left hand is then free to help stabilize the camcorder and operate the other controls. The aim is to make your shots as steady as possible.

OBJECTIVE: To minimize the effects of camera shake. *Rating* •

GRIPS

Start by getting your grip right. Hold the camcorder firmly with the grip strap tightened over your right hand. Always use your left hand to support and steady the camera – the fingers of this hand will still be close to the other operating controls. The way you position your hands will vary slightly according to the design of the camcorder, but in many cases the best support is under the lens.

PALMCORDER
Even the lightest models should be steadied with both hands. Place your left hand beneath the palmcorder's body for support.

COMPACT
Compact models with larger lenses can be supported with the fingers of the left hand, under the body of the lens mount.

TIPS

Camera shake will be magnified when using a telephoto position, so when you are shooting hand-held, use a **wide angle** as much as possible. Get into the habit of changing camera position rather than **zooming** in to telephoto.

FULL-SIZE
The full-size model is designed so that you are able to rest it on your shoulder. It can be pressed against your cheek for extra steadiness.

STABILITY

Always keep stability in mind when shooting. With all models, bolster your elbows against your chest for rigid support – although this may feel uncomfortable at first, it is well worth it for image stability. Whatever your location, look for forms of additional support. Lean firmly against a wall or brace your body against a doorway. Kneeling and sitting positions also provide stability.

BASIC STANCE
For static shots taken at eyelevel, adopt a stable stance with your feet positioned about 30cm (12in) apart. For optimum balance, keep toes slightly splayed and legs rigid.

KNEELING
Kneeling on one leg allows you to use your raised knee to prop up the arm supporting the camcorder.

FENCE
Use a wall or a stable fence for support. Sit well back against it with your feet braced. Steady the machine on your knees.

PROPPING
Use the back of a chair or a table top to prop your elbows on. When outdoors, use a low wall or car bonnet.

LYING
For steady, low angle shots, lie full-length with your elbows on the ground for support.

CUSHION •
Place a cushion under your elbows for extra comfort.

1 STANDS

Using a tripod involves a particular commitment to video-making. Since setting up the tripod takes time, you are forced to give full consideration to camera position, pictorial composition, and lens height before taking each shot. As an alternative, a monopod can be used for static shots and for when shooting in confined spaces. However, because they are lighter than tripods, they are not as stable.

TELESCOPIC •
The monopod provides basic steadiness at various heights.

STABILIZE •
Monopods can be stabilized by placing a hand on top of the camcorder and bracing the bottom of the monopod against your foot or leg.

MONOPOD
Monopods are lightweight and can be slung over your shoulder for easy carrying.

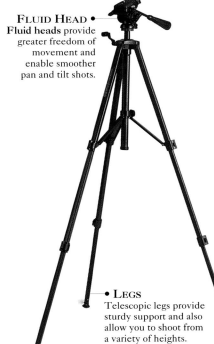

FLUID HEAD •
Fluid heads provide greater freedom of movement and enable smoother pan and tilt shots.

TRIPODS

A tripod, especially when equipped with a **fluid head**, provides excellent freedom for camera movements when following action subjects – such as sports. The maximum support the tripod offers, enables you to achieve much smoother and steadier shots.

QUICK-RELEASE
The **quick-release** mechanism allows you to swiftly dismount the camcorder from the head of the tripod, for hand-held use.

— USING A TRIPOD —

A tripod is essential for when you are using a telephoto converter and longer **focal lengths**. It is also necessary for **time-lapse**, **animation**, and **macro** shooting. Subjects that require shooting over a long period demand this type of camcorder support. Adding a tripod to your equipment not only improves stability but helps you to develop your creative skills and extend the variety of subjects you can successfuly cover.

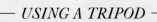

• **LEGS**
Telescopic legs provide sturdy support and also allow you to shoot from a variety of heights.

SUPPORTS

Since camera steadiness is always an important consideration, using any form of support will improve your results. With shake minimized, you are also free to use the longer **focal lengths** of the telephoto end of the **zoom** lens. The various kinds of body-mounted supports are an alternative to stands. They provide some additional stability while still allowing freedom of movement for the operator. With all forms of body-mounted supports, conventional hand-held camera movements can be executed effectively while sitting, standing, or kneeling.

HOLSTER •
Weight is taken by the shoulder and stability is provided by the brace.

BRACE •
A simple shoulder brace with a spring section provides basic stability.

SHOULDER HOLSTERS
The most effective device is a shoulder holster that provides basic stability, takes the weight of the camcorder, and gives freedom of movement for your hands and fingers on the controls. It is the ideal regular support for the heavier compact models.

CHEST PODS
A chest pod is a convenient alternative to a shoulder holster. It is strapped around the neck and helps support the weight of the camcorder. Particularly useful during long takes, it is a suitable accessory for most compact models and palmcorders.

CONNECTER •
Attach your chest pod to the base of the camcorder as you would a tripod.

BALL & SOCKET •
The ball and socket attachment allows for greater freedom of maneuverability.

ADJUSTABLE •
Adjust the pod height for added comfort and support.

SKILL

2 CAMERA MOVEMENTS

DAY 1

Definition: *Learning the techniques for the most commonly used camera movements*

PANNING AND TILTING imitate the movements we make with our head and eyes when scanning a stationary subject or following action. Camera movements, therefore, are ideal when you are unable to include an entire subject in a single, static shot – such as a panoramic view or a waterfall. They are also essential for covering most action sequences, such as children racing down the beach. Camera movements are also important for showing connections or contrast between subjects in a single shot, for instance, a shot of chimps in a cage that moves to show the crowds' reactions.

OBJECTIVE: To make your camera moves as smooth as possible. *Rating* ••

PANNING

Panning involves pivoting the camcorder on its own axis through a horizontal arc – from left to right or vice versa

COMPOSING
Compose **holds** at the beginning and end position of each pan shot. Slowly pivot around keeping the movement at a constant speed until you reach the prearranged end position of your shot.

• **ANGLE**
Avoid trying to pan too far – 90 degrees is the sensible maximum. Beyond this angle, you will create both strain and instability.

HAND HELD PANS

It is perfectly possible to carry out a successful pan with the camcorder hand-held. Ideally, the lens should be set to a **wide angle**. Hold your camcorder steady and adopt the standard position with your body facing the predetermined end position of the pan movement. Without moving your feet, twist from the waist until you are facing the start position. Once you start to shoot, untwist your body and finish in the correct position for your end shot.

KNEELING
The basic twist technique can also be used in the kneeling position. This is ideal for lower angles and for following action.

STANDING
Concentrate on stability, avoiding jerks, and keeping the camera level throughout the panning movement.

WAIST •
Steadily untwist your body from the waist as the shot proceeds.

PANNING WITH TRIPOD

Using a tripod with a good pan and tilt head will obtain much smoother results than a hand-held pan. Spread the tripod legs fully and set it up so that you are standing comfortably between a pair of the legs. Do not straddle one leg. Release the pan lock mechanism and compose the opening position of your shot. Holding the pan bar securely, gently rotate through the arc until you reach your final position and gradually come to a halt. Always avoid shuffling your feet during the move as this is likely to cause noticeable camera shake.

FOCUSING
With the right hand holding the pan handle, the left hand is free for manual focusing of the lens during the shot.

PAN HANDLE
The full movement is controlled by placing gentle pressure on the pan handle – this ensures a smooth panning movement throughout the whole shot.

• **LEGS**
The sturdy, telescopic legs of the tripod provide stability during pans.

SKILL
2

TILTING

*Tilting involves pivoting the camcorder smoothly through
a vertical arc – up or down*

SHOOTING HOLDS
As with panning, always shoot a static **hold** at
the beginning and end of any tilt shot so that
the viewer has time to register the subject
before the camera movement starts. Hold
the first image for about three seconds
before tilting – end by holding the
final image for three seconds.

BALANCE •
Tilt slowly so that
the viewer can take
in the subject and
avoid trying to tilt
too far – you could
lose your
balance.

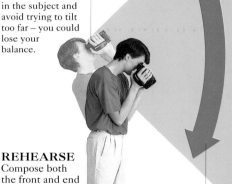

REHEARSE
Compose both
the front and end
positions of the
tilt. Whenever
possible, rehearse
the movement
before you start
shooting.

ANGLE •
Do not tilt more
than 90 degrees
otherwise you
risk straining
your back.

— PAN & TILT TIPS —

Panning and tilting can be very effective
when you know what to do, and what not
to do. Always follow these guidelines:
• With static subjects, moves should start
slowly, rise to a consistent speed, and slow
to a gentle stop. Use "buffers" to avoid
jerks at either end of the movement.
• Above all, avoid "hose piping" – that is
continuous panning and tilting across a
subject in an effort to cover it all. Instead,
break the subject into more than one shot.
• With static subjects, such as a landscape
scene, avoid panning too fast – it will only
make the shot unviewable. Judge your

speed by allowing approximately five
seconds for any object to pass from one
side of the screen to the other.
• Do not try to move too far. When hand-
holding, a natural arc of around 90 degrees
is the safe maximum angle.
• Do not stop recording while panning or
tilting – you will create an abrupt edit.
• Do not overdo camera movements.
Unless you have an action subject, no
more than one in three of your shots
should involve camera movements.
• Do not reverse pans – panning one way
and then back across the same subject.

HAND HELD TILTS

Use a **wide angle** and adopt the standard stance with the camcorder at eyelevel. Tilting up will involve arching your spine as the movement progresses and leaning back slightly. The strain on your back and your own sense of balance will warn you how far you can tilt. Tilting down involves the reverse of this procedure. By angling the viewfinder, you can tilt by cradling the camcorder at chest level and gently raise or lower it to your chest.

SUPPORTING
Use your left hand under the lens of the camcorder to provide support and to also add stability during tilt movements.

KNEELING •
For lower level tilts, kneel on one leg and use the raised knee to support your elbow at one end of the move.

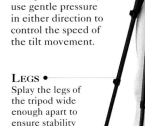

PAN HANDLE
During tilt shots, the pan handle will act as a simple lever. Only use gentle pressure in either direction to control the speed of the tilt movement.

LEGS •
Splay the legs of the tripod wide enough apart to ensure stability and allow you to stand between them with ease.

COMBINED MOVES •
If using a tripod to follow action subjects, such as a skier whizzing downhill, work with both the pan and tilt mechanisms released.

LOW ANGLES
Shoot low angle tilts by angling the viewfinder upwards and kneeling on one knee, leaving the left hand free to use manual controls.

TILTING WITH TRIPOD

The same principles of position and procedure apply here as with using a tripod for panning. Stand with your body close to the camcorder. Release the tilt lock and compose the opening position. Hold the pan bar securely and tilt by pivoting the camcorder through a vertical arc. Bring the movement to a gentle halt as you reach the end position. Use a tripod with a spirit level to ensure that the horizontal and vertical elements are straight throughout the tilt movement.

• **VIEWFINDER**
By angling the viewfinder of the camcorder, you can shoot with the tripod set at various heights.

3 MOVING WITH THE CAMERA

Definition: *Walking or moving with the camcorder hand-held*

WITH A LIGHTWEIGHT CAMCORDER in your hand and an action-packed subject, such as a children's party, your natural instinct will be to move with the camera while recording. However, to be sure of capturing the essential action you must master the technique of this first. When moving with the camera it is even more important to use the **wide angle** lens position – cutting down the inevitable shake involved in any movement.

OBJECTIVE: To help minimize on camera shake while moving. *Rating* ••

WALKING

The basic movement is to "walk" with legs bent and your body lowered. This lowers your center of gravity and enables you to avoid the rise and fall of normal walking. Make your footfalls as soft as possible. Use only short strides and keep your feet close to the ground as your legs swing forward. Walking backwards involves the same strategy, but in reverse. Be sure your path is clear before you proceed.

WALKING FORWARD
To walk forward, gently ease your toe down as the heel touches the ground. Concentrate on creating a slow-motion, gliding feeling.

CRABBING
Move sideways, arc, or **crab** around a subject, by swinging one leg in front of the other as you move. Concentrate on letting your step fall on one side of your foot, before then easing down on the other side.

— SAFETY TIPS —

Always shoot moving shots with both eyes open. You can then scan the area ahead for obstacles as well as watch the viewfinder. Avoid having your viewing eye too close to the viewfinder – you risk an eye injury if you have a sudden jolt. Check behind you for obstacles before doing shots involving walking backwards.

CRANE SHOTS

Crane shots involve either a rise or fall movement. Use them to follow action subjects or to reveal more of a scene – such as a shot of a wedding cake being cut that then rises to show the happy newly-weds. To crane up, start in a one leg, kneeling position and gently rise to full height by bracing the back leg for stability. Craning down simply involves the reverse of this procedure.

CRANE SUBJECTS
Crane shots are ideal for subjects that involve rising or falling movements. By craning, you can follow the child up the slide's ladder.

Gently rise up from a crouching position

CRANE MOVE
Ensure you are in a stable position for a steady opening to the shot – as you rise, gradually straighten your legs until you are at full height.

TRACKING SHOTS

Use a tracking vehicle or "dolly" for shots involving movement over some distance. Sit well back in it with the camcorder gripped tightly – buffered against jolts. Your weight will help to stabilize the trolley and steering, but for added stability, use your feet to brace yourself. Have someone push you smoothly along the tracking path.

TROLLEY
Make sure your companion pushes you along at a smooth and steady pace, avoiding any jolts or jerks.

— TRAVELLING SHOTS —

Shooting from cars or trains can be an excellent way of showing landscapes or city street scenes. The best shots are taken with the window rolled down but with the camcorder well inside the vehicle to avoid it being buffeted by the wind. Adopt a position with an oblique, three-quarter angle view of the passing scene. If you shoot at 90 degrees to the subject, foreground details may pass too quickly for the audience to take in.

Brace your body against the seat

SKILL

4 ZOOMING

DAY 1

Definition: *Understanding the creative potential of the* **zoom** *lens*

ALL CAMCORDERS ARE EQUIPPED with a motorized **zoom** lens controlling the **focal lengths** between the **wide angle** and telephoto position. At the touch of a button or rocker switch you can produce smooth zoom shots – zooming in to show increasingly magnified detail of a scene or zooming out to reveal a progressively wider view of the same subject. The zoom lens also allows you to take shots of various sizes from the same camera position. It enables you to frame your composition before taking a shot and also to make small adjustments during a shot to accommodate action and the movements of subjects. However, too many zoom shots will make viewing irritating – only use them for a good reason.

OBJECTIVE: Knowing when to use a **zoom** shot. *Rating* ••

• **ROCKER SWITCH**
The motor of the **zoom** is controlled by pressure on the rocker switch which activates the movement.

ZOOM CONTROL

Always be cautious about pressing the **zoom** button too often when you are shooting. Zoom shots should only be included with a true purpose as an excess of them can be tiring and discomforting for your viewers. As a general rule, not more than one in five shots in any one sequence should involve zoom movements. Make most use of the zoom button between shots rather than during them.

SHOT SIZES
Use the **zoom** to take shots of different sizes from the same camera position. You can follow the establishing shot by framing a close-up.

TELEPHOTO
Without moving the camera position at all, the telephoto position can be used to take a selection of highly detailed close-ups.

ZOOM INS

Use **zoom** in shots to emphasize an emotional moment, such as zooming in to show a new-born baby cradled in its mother's arms. You can also use a zoom in to highlight an important element within a wider shot, such as zooming in to the weight shown on the scales as the baby is weighed.

Subject and setting

EMPHASIS
The **zoom** in shot (left) "tightens" to show a close-up of the child and her pet dog, which emphasizes the special bond of affection between them. The feeling of the whole scene has been reinforced by the movement.

WIDE VIEW
In the wide view (above) the setting is established focusing on the little girl sitting with her pet dog.

Subject and setting

ZOOM OUTS

Use **zoom** out shots to disclose any new elements not evident at the start of the shot – such as beginning with a close-up of a smiling baby in a cot and then zooming out to reveal the happy parents standing over it. Alternatively, you could try involving a comic aspect by showing the baby's scattered toys.

REVELATION
By **zooming** out from the close-up (left), the whole of the setting is revealed – the child rubbing his belly with a brush. This shows the reason for his look of keen interest and impish expression – the movement has been used to tell the audience a story.

CLOSE-UP
The initial close-up (above) establishes the child and the expression he has upon his face.

4 COMPOSITION

Most **zoom** shots require that they be combined with camera movements, such as a pan or tilt move. This will produce a more sophisticated effect – enabling you to follow action or adjust the composition. It will also enable you to include new elements as you zoom out. Avoid simply "tromboning" in or out as you risk spoiling what was a well composed close-up or distance shot.

Subject and setting

SHOOTING HOLDS

When shooting **zooms**, shoot a static **hold** at the front of the zoom shot and then another at the end so that the viewer has time to register the subject matter. These holds should be about two seconds long. Do not just press the zoom control as soon as the recording starts. If you are planning to edit the tape later, make the holds about five seconds long. Then, if there is no vital action, you will be able to edit the front and end of the shot together as separate shots without the zoom move.

OPENING SHOT
With any **zoom** shot, always hold the opening position long enough to enable your audience to take in exactly what is being shown.

MANUAL ZOOMING

On most camcorders, the speed of the **zoom** movement is fixed by the motor. Some offer two speeds of motorized zoom and the more sophisticated models enable you to vary the speed of the zoom move. However, the governed speed can often be a problem, especially with fast moving action subjects. Machines with manual zoom control enable you to override this.

ZOOM LEVER •
The manual **zoom** lever is attached to the barrel of the lens.

CONTROL
Gently hold the manual **zoom** lever between just the tips of your thumb and first finger to control the speeds of your zooms.

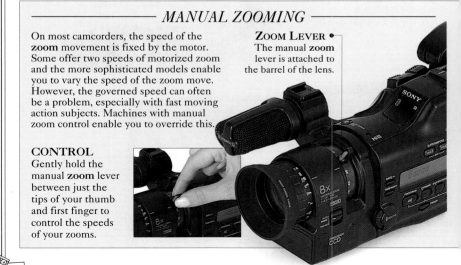

UNBALANCED
Just **zooming** out leaves too much space at the top of the screen and an incomplete subject.

COMPOSITION
Left: Tilting down while **zooming** out from a close-up of the little boy, has produced a final composition that is evenly balanced. There is natural headroom at the top of the screen and all the toys on the floor are brought into view.

ZOOMING OUT
The **zoom** out has revealed more of the scene. This shot is gradually revealing what the little girl is unwrapping. With static subjects (such as buildings and landscapes, or where there is no significant change in the scene), the zoom section of the shot could be edited out later so that just the front and end **holds** are cut together.

END POSITION
The end position of the **zoom** should be held long enough for the audience to take in the scene or for the action to be completed.

VARYING ZOOM SPEEDS
Manual control of the **zoom** enables you to change the **focal length** of the lens more rapidly between shots. It also allows you to vary the speed of zoom movements. For instance, you can execute powerful "crash" zooms with a rapid movement of the zoom lever – ideal for simulating shock or surprise. Remember that manual zoom movements are likely to be less smooth than motorized ones. Always try to practice before recording. With action subjects, you can easily adjust your shot size to suit the speed of the movement, such as with the skier coming towards you, shown in the three pictures below.

Telephoto view

Zooming out (with action)   *End of zoom*

SKILL

5 SOUND SENSE

Definition: *Understanding sound as part of the video recording*

SOUND IS AN ESSENTIAL ELEMENT in video-making. The keys to recording "good" sound are being aware of the capabilities of your camcorder's own audio system and the techniques for getting the best results. Most camcorders are equipped with a simple omni-directional mike – that is, one that picks up sound from all around. Unfortunately, it also records any handling noise you make while operating the controls with your fingers. In quiet situations, it may also pick up the sound of the **zoom** and auto-focus motors.

OBJECTIVE: To record sound clearly in a variety of situations. *Rating* ••

BASIC SOUND RECORDING

Every time you press the record trigger, you automatically record sound along with the pictures

• **BUILT-IN MIKE**
When using the built-in mike to record speech, ensure you are close enough to the subject to record distinctly.

ALC

All camcorders come equipped with an automatic level control (**ALC**). The audio circuitry adjusts the sound level to produce a recording of even volume. This helps to smooth out any sudden changes in sound without you having to make adjustments. Do be aware that in very quiet situations the ALC will overcompensate by raising the level of background sound. A few camcorders allow manual setting of the sound level using a volume meter.

CONTROL BUTTONS
Time spent learning the position of the control buttons until you can touch them instinctively will enable you to reduce, or even eliminate, any handling noise.

HEADPHONES
When possible, use headphones to monitor the quality of your sound recordings. The closed-back headphones are especially effective.

MIKE TEST

Check the effectiveness of your camcorder mike with this simple test. Have someone read from a book while you record them at various distances from your camcorder. Start at one meter (3ft) away and then move away, recording in one meter (3ft) steps. Play back the results to confirm the maximum distance for audibility. You can also monitor the effect of the **ALC** – as you move further away, the reader's voice will become less audible and the surrounding background sound will increase in volume.

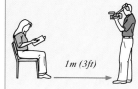

1m (3ft) *1m (3ft)* *1m (3ft)*

DISTANCE
Make a note of the maximum distance you can be from the reader and still hear the words distinctly.

INDOORS

When shooting indoors, be aware of the effect of acoustics on the clarity of sound. When recording, move the camcorder as close as possible to the source of sound for improved results. This may mean having to use **wider angles** and adjusting your shots. The best acoustics are often to be found in a domestic living room with a mixture of sound reflective materials (walls, glazing, and wood) and those that are sound absorbent (carpets, curtains, and soft furnishings). Avoid recording in bathrooms, echoey corridors, and the corners of rooms when possible.

ACOUSTICS
Choose your position carefully, especially when recording in large halls, churches, or any rooms with bare walls and glass. The hard surfaces reflect sound and speech, making sound reverberate and become indistinct.

OUTDOORS

To avoid problems with wind rattle in the mike, either buffer the wind by standing with your back to it or screen the mike with a natural windbreak such as a wall. Extraneous background sounds, such as a jet overhead, can be reduced by using an auxiliary directional mike.

SHIELD THE MIKE
When shooting outdoors, always look for an opportunity to shield the mike from wind noise by standing behind something. You may also be able to use this for support.

WINDSHIELDS •
Foam windshields are essential for mikes when recording in windy conditions.

SKILL
5

AUXILIARY MIKES

*You can overcome the limitations of built-in mikes
with various specialized auxiliary mikes*

MOUNTED MIKES
Mounted mikes plug into the camcorder's
sound socket and override the built-in mike.
They can reduce or eliminate handling and
motor noise problems although their greater
sensitivity means that care is still needed.

ACCESSORY MIKE
Once you have learned the basic skills of sound
recording, you can then start to extend your
creative potential with various auxiliary mikes.

CARDIOD MIKES
Cardiod mikes pick up sound
mainly from within an arc of
180–200 degrees around their
axis with some sound coming
from behind. Their greater
directionality means
that speech can be
recorded up to
3½m (12ft) from
the speaker.

*Polar response for
cardiod mike is
180–200 degrees*

INTERVIEWS
A hand-held **cardiod** mike is ideal for a two
person interview situation. The mike can be
held halfway between the two speakers.

SUPERCARDIOD MIKES
Supercardiods cover a narrow angle of
acceptance of around 120 degrees.
Any sound outside this angle is
largely rejected, although not
eliminated. These mikes can
successfully pick up sound
over greater distances so
are ideal for use when
you are unable to
get close to your
subject.

DIRECTIONAL USES
Supercardiod mikes are useful outdoors in
situations where you want the sound to
match the impression of your telephoto shots.

*Polar response of
supercardiod is
120 degrees*

HYPERCARDIOD MIKES

Hypercardiod ("shotgun") mikes
favored by professionals
feature the narrowest
acceptance angle
of all mikes.

*Polar response of
hypercardiod is less
than 90 degrees*

SPECIALIZED SUBJECTS

Hypercardiod mikes are highly effective for recording over distances, making it ideal for recording subjects, such as wildlife and birds.

TIE-CLIP MIKES

Tie-clip mikes are small units that are ideal for interview situations. The mike is clipped to the speaker's clothing, above chest level and the cable is run to the camcorder's mike socket. The cable can be made less obtrusive by hiding it under the clothing. This mike is usually used when only the interviewee's answers are to be recorded.

*Tie-clip mikes can
accept sound from
360 degrees*

PRESENTERS

Small clip mikes can be used by presenters or interviewees. The wearer must take care not to brush against the mike itself.

MIKE TESTS

ANGLE OF ACCEPTANCE

Test the angle of acceptance of any auxiliary **cardiod** mike by recording someone reading out loud about 2m (6ft) away from the mike. The effect of the automatic level control (**ALC**) will be to raise the general level of the background sounds as the reader moves outside the immediate area of the polar response pattern. This will result in the voice becoming both less audible and distinct.

Camcorders with manual sound level controls allow you to conduct a purer test. Fix the sound input level for the first position – this will enable you to observe the needle on the volume meter showing a steadily lower level of recorded sound as the reader moves around the arc.

-2 dB   *-10 dB*   *-23 dB*

2m (6ft)

*Test the polar response of an auxiliary mike
with the camcorder pointing in a fixed direction*

6 USING LIGHTS

Definition: *Ensuring that subjects are effectively illuminated*

CAMCORDERS CAN RECORD IMAGES in low light – down to five lux and less (see pp.18-19), although under these conditions the definition will be poor and colors subdued. Camcorders really only produce good quality pictures at around 1,000 lux and above. Shooting outdoors during daylight hours, even on an overcast day, always produces good results. Shooting indoors in a room with plenty of sunlight can also be successful – particularly where there is an even spread of light. However, once daylight levels fall, or you need to shoot at night indoors, you will need to consider supplementary lighting to maintain picture quality.

OBJECTIVE: To choose and use lights to the best effect. *Rating* ••••

• LIGHTS
On-board lights draw their power from the camcorder or a separate re-chargeable battery and produce a narrow beam of light over a limited area.

ON-BOARD LIGHTS

Small, low wattage, tungsten-quartz lights mounted on the camcorder accessory shoe, are the simplest to use. Use them for close shots – 2-3½m (6-12ft) from the subject, making sure that you compose within the area covered by the light. They are ideal for shooting parties, celebrations, and family events. Battery life is limited to less than twenty minutes, so keep plenty of spares to hand.

FILL LIGHT
Battery lights can be used to provide "fill light" in shadow areas and to overcome backlight problems (see pp.18-19) by providing frontal illumination. They are ideal for close shots, adding sparkle and lifting the general level of illumination.

HAND HELD LIGHTS

Hand-held lights with separate power packs are generally more powerful than on-board lights but require a companion to hold them. Keep them about 30 to 45 degrees away from the axis of the camera to provide "modeling" on the subject. Hold them above the height of the lens and angle them downward.

Hand-held light •

• Power pack

SMALL GROUPS
Hand-held lights are capable of providing the main light source over a wider area for full-length shots of people or small groups.

POWER PACKS
These heavy duty battery packs are worn over the shoulder and will provide power for twenty to forty minutes, but may take several hours to recharge from an outlet.

ELECTRIC LIGHTS

Electrically powered video lights produce a strong flood of light over a wide area – providing a good level of illumination. With some video lights, you are able to adjust the spread of light from a wide "flood" position to a narrow "spot". They may also have adjustable "barn doors" – metal flaps for controlling the spill of light.

• Power is usually between 500 and 2,000 watts

Each light has • its own telescopic tripod stand

EVEN ILLUMINATION
Set lights high, at about 45 degrees to camera axis and far enough apart to cover the area.

BACKLIGHT
A backlight set high behind the subject will separate figures from the background scene.

SKILL

6 MIXED LIGHTING

As the **color temperature** of daylight is lower than that of tungsten lighting, there will be an orange tinge to those parts of the picture lit by the light if you have set the white balance to the daylight setting. However, a **blue gel** spread over the light, will raise the color temperature to that of daylight.

FLUORESCENT LIGHTING
Above: Overhead fluorescent strip lights produce an even spread of diffused light that is ideal for basic illumination. However, it has a high **color temperature** that produces a green cast to the scene when used with the tungsten setting. To compensate, try using the AWB or the manual white balance lock.

PHOTOFLOODS
Left: To light large areas, use high output, photoflood bulbs in place of domestic, tungsten bulbs – remove shades for safety.

REFLECTED LIGHTING

Free-standing and hand-held video lights can produce harsh illumination, especially if used close to the subject. They also tend to produce a "hot spot" of bright illumination at the center of the beam. To produce even lighting of a lower contrast which is favored by video, try "bouncing" the light by reflecting it off white painted surfaces.

DIFFUSING LIGHT

DIFFUSER
A sheet of diffuser can be pegged to the barn doors and bowed away from the light to provide a more even spread of illumination.

By diffusing light you can produce an ideal, low contrast, even illumination. Some video lights incorporate built-in diffusers, but diffusion works best when there is some space between the light and the diffuser material. Use diffusers by clipping them to the lights' barn doors, by holding them in front of the lights, or by fixing them into frames.

LIGHTING ACTION

With a party, dancing, or indoor games involving movement, direct lighting is not as effective and creates too many shadows. To illuminate your setting successfully, try spreading the lights as wide apart as possible, setting them high and bounce the beam of light off walls or a reflective (white painted or light toned) ceiling. You can still use a separate, hand-held or on-board, battery light for close shots.

IMPROVISED LIGHTING

When shooting indoors at night, for example at a family gathering, the main concern will be to raise the level of illumination to ensure satisfactory picture quality, good definition, bright accurate colors, and to increase depth of field. If electric or battery operated video lights are not available, you can improvise by using the existing domestic facilities while still keeping the atmosphere of the occasion.

DOORS •
Open doors to allow light to "spill" from adjacent areas.

SHADES & LIGHTS •
Turn on all the available lights, remove any light shades, and redirect the spotlights as necessary.

• DESK LIGHTS
Bring in additional domestic lights, such as desk lights. These can be angled to illuminate a particular area.

• BULBS
Change existing bulbs to those of a higher wattage or replace with photoflood bulbs.

SAFETY TIPS

• Photoflood bulbs become very hot in use so do not cover them in any way – shades could easily melt and burn.
• Always handle video lights with care and use garden gloves to adjust the barn doors when they are hot. Remember to turn the lights off when moving them to preserve bulb life and allow the units to cool down before packing them away.
• Coil light cables loosely and tape them

securely to the floor whenever possible, but especially when children are about.
• Weight down the base of your tripod to prevent accidental tip-ups.
• Take great care not to overload domestic lighting or main power circuits. As a guide to power availability, simply multiply the amperage by the voltage. The result will determine the maximum wattage you can use for each individual socket or circuit.

SKILL

7 COMPOSITION

Definition: *The arrangement of the visual elements within a shot*

WHATEVER THE SUBJECT, the classic concerns of composition, including balance, perspective, shape, and form, apply as much to the framing of a video shot as they do to photography and painting. Each video shot must be considered in relation to the one following it, therefore, creating visual variety is an important element.

OBJECTIVE: To maximize the effectiveness of your shots. *Rating* ••

STATIONARY SUBJECTS

Guidelines to composing static shots of subjects such as landscapes, buildings, and street scenes

Subject and setting

FRAMING

As a video-maker, you must compose a scene within the fixed format of the TV screen – a proportion of 4:3. However, even within this restriction, you can create visual variation by, for instance, framing through doorways, windows, or using foreground objects.

RULE OF THIRDS
Achieve picture balance by organizing the composition so that key elements – such as the horizon or verticals, fall along lines dividing the scene into thirds. Do not place subjects in the middle of the frame simply because they are important.

Landscape makes up two thirds

Landscape makes up one third

FRAME IN A FRAME
Use a vignette mask fixed to the lens with an adaptor (far left) to create any screen shape you wish. Doorways (left) and windows can also be used effectively to create a "frame within a frame".

VIEWPOINT

Use a viewpoint in which the main elements are presented to create a sense of depth. Shoot buildings and streets from a three-quarters angle to create strong diagonals and position yourself so that people move towards the camera, giving a sense of space. Using foreground elements will help enhance the illusion of depth. Use focus differentials to emphasize the important elements and the contrast of light and shadow to create strong shapes within the frame.

VARY HEIGHT
Vary your camera height and compositions – try kneeling, lying on the floor, standing on a chair, or operating with the camcorder above your head.

VARY ANGLES
Try shooting scenes from a variety of angles rather than staying with just the single camera position.

VARY FOCAL LENGTH

Use the full range of **focal lengths** on your **zoom** lens. Compose in telephoto for striking effects with compressed perspective and to exploit narrow depth of field. Shoot in **wide angle**, using a lens converter for dramatic composition with great depth of field. Vary your shot sizes, and move to your subject. Close-ups look best on the television screen; big close-ups have extra impact. Examine the examples below to see the effect of varying the focal length. In a wide angle, the subject is lost against the background, but by moving further away you can use the telephoto effect to suppress the background. Extreme telephoto will produce the greatest impact.

Wide angle *view*

Background out of focus

Full telephoto effect

SKILL

7 FRAMING & MOVEMENT

*Video creates certain compositional demands, particularly
in relation to moving subjects*

WALKING ROOM

When panning moving subjects
such as people, compose your
shot so that there is more space
in front of them than behind,
otherwise the person appears to
push the edge of the frame along.
Start panning as the subject enters
the frame and continue at a constant
speed. To end the shot, stop panning
and let the subject move off-screen.

MOVEMENT IN THE FRAME
For static shots of moving subjects, be sure
you compose loosely enough to include the
subject, but not distracting surroundings.

*Subject pushing
the frame along*

*Subject has space
to move into*

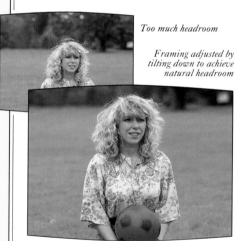

Too much headroom

*Framing adjusted by
tilting down to achieve
natural headroom*

HEADROOM

When composing full-length shots of
people, avoid placing their heads in
the center of the frame – this creates
an unbalanced composition with too
much space above them. Instead, tilt
down from this position to produce a
more natural result with only a small
amount of space between the person's
head and the top of the frame.

MOVEMENT OF THE FRAME
With active subjects, such as a person
playing "catch ball", be prepared to adjust
the composition and make small camera
moves to accommodate changes in position.

COMPOSITION TIPS

• When panning a moving subject, keep
it one third in from the edge of the frame.
• When tilting to shoot a subject that is
moving upwards, keep it approximately
one third above the bottom of the frame.
• When tilting downwards, keep two
thirds of the frame below your subject.

• With long shots, allow some headroom,
otherwise the figure will look crushed by
the top of the frame.
• Avoid having the subject's chin resting
on the bottom of the frame.
• Compose close-ups so that the person's
eyes are a third below the top of the frame.

LOOKING SPACE

Static shots of people always need careful, spatial composing. Arrange your shot framing so that there is more screen room in the direction the person is looking, than behind them. This produces a composition that is both comfortable and logical. It also helps to reinforce the direction of gaze, of a profile, or three-quarter view of your subject.

LOOKING OFF-SCREEN
With the subject looking off-screen, either in profile or a three-quarter view, always leave more space in front of them than behind.

No looking space

The screen space reinforces girl's direction of gaze

A long shot

A mid-shot

A close-up

A big close-up

SHOT SIZES

As many of your shots will include people, it is a good idea to learn the standard compositions. These are the shot sizes that feel most comfortable for the viewer: • The long shot (LS) in which the figure is seen full-length. • The mid-shot (MS) with the bottom of the frame cutting off just below the subject's waist level. • The close-up (CU) that cuts through at upper chest level. • The big close-up (BCU) which concentrates on the composition within the face – usually just cutting through both the subject's chin and the forehead. Close-up shots are particularly effective when shown on the TV screen – they bestow a feeling of intimacy with the subjects.

COMPOSING
Each time you set up a shot, be it any of the above, carefully compose the elements so as to frame your subjects most effectively.

AVOID

• Avoid framing shots that cut people off at the joints – you risk creating disembodied heads or footless figures.
• Don't use a background that is full of detail when composing a shot of people.
• When using monochrome viewfinders, check the background color to ensure the figures do not become lost against it.

8

A SENSE OF DIRECTION

Definition: *Creating a credible continuity of events the audience will understand*

THERE ARE NO STRICT RULES to video-making, but, as with language, there is a recognized "grammar". The syntax of video evolved in the early years of cinema and has been refined since then through the practices of television. What this grammar provides is a set of conventions that enable your audience to make sense of what they see on the screen. You do not need to follow these guides "to the book", but in order to produce fully comprehensive videos, it is essential that you are aware of how, and where, they can help you.

OBJECTIVE: To retain a sense of direction from shot to shot. *Rating* •

LINE OF ACTION

To preserve continuity of screen direction, shoot consecutive shots from one side of an imaginary "line of action". If the action is to make full sense, a shot establishing a subject's direction of movement (such as left to right across the screen) needs to be followed by a shot showing the move in the same direction.

A   START
A – Here, the race track is the line of action. This shot shows the direction of movement.

SAME DIRECTION
B – By staying on the same side of the line, the second shot confirms the direction.

SCREEN DIRECTION

If a shot of the bride's car leaving her home shows the vehicle moving out of shot to the right of screen, the following shot of it on its journey (to the church) should show it enter the screen, left. If you get this the wrong way around, your audience may get the impression that the vehicle has got lost, or worse, the bride has decided to return home. Remember – out right, in left, and vice versa. If the bridal car really has got lost, you will need to take a shot of the vehicle turning around in the roadway before another showing it arriving at the church from the new screen direction.

WRONG WAY
If you show the vehicle entering the screen from the wrong side, your audience may lose all sense of screen direction.

RIGHT WAY
By retaining a sense of screen direction, you are able to link shots taken at different locations in a continuous sequence.

REVERSING

Overcome the problem of "reversing" screen direction by using a neutral shot taken along the line of action. This will result in there being no left or right screen direction. This type of shot can be used effectively as a bridge between two shots taken from either side of the line of action.

E

C

D

NOT CROSSING
C – The sense of direction will continue as long as you avoid "crossing" the line.

NEUTRAL SHOT
D – To bridge a change in the screen direction, insert a shot that has no apparent direction.

NEW DIRECTION
E – The next shot taken from the opposite side of the track will establish a new direction.

9 SPACE & TIME

DAY 1

Definition: *The relationship of events in succeeding shots*

VIDEO PROGRAMS ARE COMPOSED of numerous, separate shots. It is therefore essential that you ensure the audience is aware of the spatial and temporal relationship between the events shown in them. In most instances, spatial relationships between people are easily shown in initial long shots. However, once you move in to take closer shots of individual subjects, you need to retain the sense of their spatial relationship in order not to confuse your audience.

OBJECTIVE: Creating relationships within sequences of shots. *Rating* •

A SENSE OF SPACE

A guide to the spatial conventions to be aware of and how to follow them whenever possible

EYELINES

Any two successive shots of people talking or relating to each other, must show them looking in opposite screen directions. Imagine an eyeline along the axis of their gaze. The first and second shots must be taken from the same side of this line. The key rule is shoot from just one side of the eyeline.

Subject and setting

CORRECT EYELINE
Shooting from just one side of the eyeline with the people looking in opposite screen directions, will make them appear to be relating to one another.

INCORRECT EYELINE
Crossing the eyeline between shots creates the impression that the two speakers are ignoring each other.

REVERSE ANGLES

Always make sure that **reverse angles** match each other. The principle of eyelines applies to both inanimate and live subjects. For example, a close-up of a person reading a notice, will need to be followed by a shot of the notice taken from the same side of the eyeline between the person and object.

Subject and setting

CORRECT ANGLE
Even if you include both figures, always shoot successive shots from one side of the eyeline to avoid disrupting or confusing your viewers.

INCORRECT ANGLE
Moving diagonally across the table, crosses the eyeline. Shooting from just one side of the players prevents this.

EYELEVEL
A close-up of a child with a picture book is best shot from a complementary low angle.

COMPLEMENTARY

The spatial relationships between subjects are reinforced by using "complementary" angles. With a subject, such as a child reading a picture book, the close-up of the child should ideally be taken from a low angle and complemented by a following shot of the book as seen from the child's eyelevel.

MANIPULATING REALITY

The conventions of screen relationships can give great freedom to manipulate reality. A close-up of someone looking or pointing off-screen in a particular direction, makes the audience assume that any succeeding shot from a complementary angle reveals what the person was looking at, even if it is not true. These conventions of eyelines and spatial relationships can be seen at work in most dialogue scenes in movies.

Complementary angle of subject

Girl looking off-screen at subject

A SENSE OF TIME

*Getting a sense of appropriate shot lengths by developing
your sense of screen time and the skill of editing*

SHOT LENGTHS

Video-making necessitates that you
develop a sense of screen time and
the skills to manipulate it. Each time
you press the "record" button, think
about the length of the shot. With
static shots of stationary subjects, shot
length is determined by the amount
of information the audience needs to
take in. In general, the wider the shot
and the more detail it shows, the
longer it needs to be on screen. As a
guide, allow a minimum of three
seconds from the start of recording.

ESTABLISHING SHOT
Shots of new subjects need to be held for
about seven seconds, but this may be less
or more depending on the amount of detail.

LONG SHOT
Long shots of one subject
with less detail should be
held for around five seconds.

MID-SHOT
A closer shot can be shorter
still unless there is some
significant detail or action.

CLOSE-UP SHOT
Close shots showing details
of the wider scene should be
held for three to four seconds.

*Hold continuous
activity of bustling
market place*

*Show repeated
cycles of action
of girl swinging*

CONTINUOUS ACTION
With scenes of repeating cycles of action –
hold long enough for the cycle to repeat (at
least once) and cut at the end of the cycle.

ACTION SUBJECTS

Any static shots involving continuous
action, activity, or slow movement – a
view of a market scene – will need to
be up to half as long again. Where you
are following action with camera
movements, it is important to let the
camcorder settle momentarily at the
end of the shot before cutting. For
shots that involve words which the
audience must read, allow about three
words per second, then add one third
of the total for slow readers.

CUTTING POINTS

Follow the skier across the screen

Cut after skier has left screen

Much of the time, cutting points and shot lengths will be determined by the action you are recording and the degree to which the subject will hold the interest of your audience. With static or panning shots of action subjects, such as a skier traversing downhill, the logic is simple – cut once the subject has left the frame. It is often essential when shooting activity subjects that the viewers see the action from start to end. They will feel cheated if you cut too soon.

Cutting here is too soon

Cutting later will show the complete action

COMPLETE ACTION

In the above pictures you can see the skier going through the waves. With a shot like this, follow the action until the skier has left the screen, then cut. Holding the shot of the empty frame would lessen the impact of the shot.

CUTTING TOO SOON

In this shot (far left), the video-maker has cut too soon for the audience to see the action completed. However, in the second shot (left), the video-maker has continued shooting long enough to see the child actually get on the horse and begin to ride off – showing the whole, completed action.

SOUND & PROGRAM LENGTH

Timing has to do with the length of individual shots and the overall length of the program you make. Often, shot length is determined by the sound you are recording, so keep listening. A guide's description of a place of interest or the best man's speech at a wedding could be spoiled if the audience does not hear the essential parts. Program lengths will vary according to the subject matter, but you should always keep your audience in mind when deciding how much to shoot. Weddings and sports events may take a fixed time, but do not necessarily demand that they subsequently be seen in their entirety. If your audience get bored, you will lose their attention, so it is always better to be brief.

The sound that is being recorded will often determine how long the shot is held for

SKILL

10 EDITING

DAY 2

Definition: *The order and length of shots in a program*

EDITING IS THE SKILL that lies at the heart of video-making. In its simplest form, it refers to the order and length of shots in a program, at its most creative, it will determine your audiences' response to the subject. Editing in-camera is the simplest way to edit since it is automatically carried out at the time of shooting. Your video will consist of a sequence of linked shots – shown in the precise order you shot them. Therefore, the length of the shots and program will be fixed. More advanced tape-to-tape editing is achieved by copying your original material onto a second tape using an Edit VCR. Edit controllers can be used to provide more accuracy.

OBJECTIVE: To create sequences of images that make sense. *Rating* •••••

CUT-AWAYS

Use **cut-aways** to abbreviate screen time. A cut-away is a shot, related but separate to the main subject, that can be used to link two periods in time. Try using people watching an event to make a cut-away. At a barbecue, use a cut-away of the cook to cover the time it takes to grill a steak. Two or three cut-aways together are useful as they suggest longer periods of time.

Subject and setting

CLOSE-UP
Start with a close-up shot of the child sitting on the beach, starting to build a sandcastle.

CUT-AWAY
Follow the close-up with a **cut-away** of the encroaching tide – away from the castle.

FINAL SHOT
Cutting back for the last shot shows the child with her now completed sandcastle.

CUT-INS

Cut-ins are useful when the event you are recording is too long to show in its entirety. You can create the impression that time has passed by cutting-in from a wider shot to show close details of the scene then return to the wider view. At a dinner, use cut-ins of the people eating or talking to link the different stages in a meal seen in wide shot. Avoid turning the camcorder off and on while keeping it in the same position, otherwise you will create distracting **jump cuts**.

— BACKSPACE EDITS —

Each time you "pause" the recording, the camcorder's tape rewinds over the last two seconds of the shot you have just recorded. When you press "record" again, the tape moves over the same section and starts to record about half a second before the end of your previous shot. This backspacing facility ensures that the point at which the shot changes is a **clean edit**. However, each time you press the record button, the shot starts about 1½ seconds later and wipes the last ½ second of the previous shot, so take account of backspacing when you are calculating the shot timings and the moment to start recording a scene.

LONG SHOT
Long shot of mother and her children sitting outside a cafe having lunch together.

CUT-IN
To show time passing, **cut-in** to a detail of the girl now eating a piece of cake.

LAST SHOT
The waitress starts to clear the table now the family has finished their meal.

FADES & DIGITAL TRANSITIONS

Many camcorders are now equipped with faders. Fade-outs and fade-ins can be used to overcome the problem of **jump cuts** and indicate the passage of time. They are also particularly appropriate as a link between different sequences and subjects. Some camcorders can create additional complex transitions using digital effects. The end of one shot is "frozen" as a static image that is stored in the electronic memory. This is then mixed to the next scene; the frozen image gradually dissolves to show the new live action shot. Alternatively, the frozen image can be replaced by the new shot using a wipe – the new live action shot is progressively revealed by simply wiping smoothly across the screen from one side to the other. Wipes are an effective way of creating a visual transition to link related subjects, but should not be used too often.

Frozen image of outgoing scene

New scene wiped across screen

Shot of new location established

10   TAPE-TO-TAPE EDITING

Copying material from the original tape, shot on the camcorder, onto a second edit master tape

PREPARATION

In preparation for editing, log your tapes by listing the shots and the start point on the camcorder's tape counter. Plan the edit on paper before you start and use this shot list as a reference when you are searching through your tapes for the required material. The primary equipment you will need is a VCR with an edit facility to ensure **clean edits**, your TV set, and an appropriate set of connection leads.

EDIT TAPE

Place a blank edit tape into the VCR (with the input set to auxiliary), run it for about 30 seconds – as a protective leader – then press pause/record button.

• *Pause*

Record •

2. PAUSE/RECORD

Both pause buttons are then released at the same time and the VCR will start to record the selected section. Watch the scene on the TV monitor for both the picture and sound quality during the copying process. Keep your finger poised over the pause button.

• *Pause*

1. CAMCORDER

Once the edit tape has been loaded into the VCR and set to pause/record, the original material is then played on the camcorder. Run down the tape until you locate the "edit in" point of the first shot you require and then put the machine into the pause mode.

EDIT CONTROLLERS

Using an edit controller gives maximum control over the editing process, enables assemble and insert edits to be performed, and provides the greatest degree of editing accuracy. A single set of controls is used to command the transport functions of both the player and recording machines. Most importantly, edit controllers enable you to preview a rehearsal of each edit before you start to record (and change the edit point if you are dissatisfied), or create a sequence of edits from a single tape that can later be altered. The ideal set-up uses a basic VCR as a player, linked via the edit controller to a VCR. Each VCR has its own TV or monitor so that the shots can be visually compared at their edit points. Some of the more sophisticated edit controllers also incorporate a form of **time-code** system.

• **MONITOR**
The TV should be switched to the usual video channel. It will then act as a monitor for both the camcorder and the VCR.

• *Pause (VCR)*
Stop (camera) •

3. PAUSE/STOP

As soon as you reach the "edit out" point on the selected shot, press the pause button on the VCR and stop the camcorder. You can then either check on the VCR the edit you have just made or find the next shot along on the camcorder and repeat the process.

4. JOG SHUTTLE

Sophisticated VCRs with jog/shuttle controls can help with accurately locating edit points.

TAPE-TO-TAPE TIPS

To make assemble editing more precise, take into account the delay in the start of the recording when you release the pause button of the VCR – about two seconds long. Allow for this delay by predicting the start or "in" point on your original material and allowing a little extra time at the end of each shot that you edit onto the VCR. To ensure a "clean" start to your edit tape, use one that has been "blacked" first. Record the section at the front of the edit tape using your camcorder with the lens cap on. This will ensure that you start with a black screen and have the necessary **sync pulses** for a stable first edit point.

INSERT EDITING

Many camcorders have an "insert editing" facility that enables you to arrange new shots at specific places in your recording. Insert editing is also possible as part of the tape-to-tape editing process, giving you greater freedom to place additional shots after assemble editing. This system enables you to add close-up details of subjects already covered and insert titles and captions. With some models, both the picture and sound are replaced, with others, it is only the picture with the original sound recording continuing to play over the edit points.

SHOOT TO EDIT

Tape-to-tape editing lets you "shoot to edit". In this sequence, the first, fourth, and sixth images are taken as a single pan shot when the engine passes by. The second and fifth are one shot, taken later and from a different position. The last shot is taken from another place. The three shots cut together make a sequence in which the engine appears to pass us by only once.

SKILL

11 ROSTRUM SHOOTING

DAY 2

Definition: *Shooting two dimensional subjects and copying onto video*

YOU DON'T ALWAYS NEED live action to have a worthwhile subject to record on video. There are good reasons to incorporate photographs, slides, or illustrations as part of a program – wedding videos can benefit greatly by the inclusion of some of the "official" photographs, and holiday videos can be enhanced by using stills and maps. You can also use paintings and artworks as well as home drawn titles for your programs. The ideal way to shoot all these items involves some form of "rostrum" set-up.

OBJECTIVE: To ensure the best reproduction of images. *Rating* •••

TAKING STILLS

Set up the materials to be used, either vertically, by pinning to the wall or a board, or horizontally, by lying them flat on a table or the floor. For best results and color reproduction, use a small on-board battery light (to cover the area of your subject) or a pair of electric video lights set at an angle of 45 degrees to the subject to provide an even spread of light across the area. Alternatively, improvise by using two desk lights – these can be fitted with photoflood bulbs for extra brightness.

FOCUS
To ensure correct focusing, position the camcorder far enough away from the subject for it to be outside its "close focus" distance, this is usually about one meter (3ft) away.

• REFLECTIONS
Pictures behind glass are likely to produce reflections, including that of the camcorder. Tilt the picture slightly or use a polarizing filter to eliminate unwanted reflections.

TRIPOD •
Use a perfectly leveled tripod to ensure that the image is carefully aligned. Adjust the central column until you are sure the lens is directly opposite the center of the subject.

CAMERA MOVEMENTS

Take great care with camera movements when working on a small scale. Connect the camcorder to the TV so that you can monitor your results and retake shots if necessary.

• CAMCORDER
Set the camcorder up in a dark room, on a tripod, close beside the projector – avoid visual distortion by aligning it along the aim of the beam.

PROJECTED SLIDES

Project slides onto a screen in the normal way. Make sure the image is bright enough to get a good **exposure** and large enough to allow for camera movements. Take care when panning or tilting – rehearse the movement before you record the shot. Set the white balance to the tungsten setting and check the image quality by monitoring the picture on the TV. When framing, take account of the projected image shape – particularly with "portrait" slides – and compose the shots accordingly.

• Result on monitor

COMPOSITION

Remember that the video image will not be the same shape as your photographs. Composing will mean framing parts of the picture and then **zooming** and moving as necessary. Still images are easier for an audience to "read" so your shot timings can be slightly shorter than those used for live action.

• Reframed composition

• Portrait shaped image

— MOVIE FILM —

Home movies can be copied onto video in a similar manner to video slides. Keep the projected image relatively small to ensure that it is as bright as possible. The film image is the same ratio as video so there should be no loss at the edges of the frame. If possible, use manual iris setting to set an average **exposure** for each film – otherwise auto-exposure changes may be evident as the film is run. With silent movie films, problems with image flicker caused by the different frame rate of film and video can be reduced by adjusting the playing speed of the projector. With sound films, use an audio lead from the projector output to the camcorder "line" input if it has one. Alternatively, place the auxiliary mike in front of the projector's loudspeaker.

SKILL

12 ANIMATION

DAY 2

Definition: *Creating the appearance of movement*

MANY CAMCORDERS incorporate **interval timers** and **animation** facilities that enable you to record very short shots – about one sixth to one half of a second – of only a few frames duration. Although this does not compare with the single frame recording possible with movie cameras – and which is the basis for creating fully animated film and TV cartoons – you can still create fascinating animation.

OBJECTIVE: To create **animation** using an **interval timer**. *Rating* •••

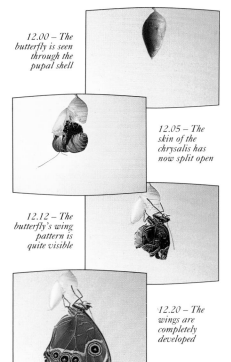

12.00 – The butterfly is seen through the pupal shell

12.05 – The skin of the chrysalis has now split open

12.12 – The butterfly's wing pattern is quite visible

12.20 – The wings are completely developed

INTERVAL TIMER
With the **interval timer** set to record 12 frames every 30 seconds, you can compress a 20 minute event into 20 seconds recording.

TIME-LAPSE

Use **interval timers** to record lengthy events. These will appear speeded up when played back. Where there is minimal change in subject position between each burst of recording, such as a flower blossoming, the effect will be a smooth, continuous movement. For **time-lapse** work, use a tripod and mains power, and ensure any lights required are fitted with new bulbs. Compose carefully to ensure that the action you want is all within the frame.

PIXILATION

Time-lapse facilities can also be used to **animate** humans. Set up a long shot of a person standing with their feet together. After each burst of recording, have them move one foot in front of the other (heel to toe), draw up the other beside it, and then adopt the same pose. Repeat until they have moved past the camera. They will appear to slide along.

Set the interval timer to allow the person enough time to make each move

ANIMATION FACILITY

Camcorders with an **animation** facility enable you to record just a few frames (4–8) each time you press the button. Basically, animation is achieved by making very small changes in the pose of your subject matter between each recording, although your results will still be a series of "jerky" movements.

TRICK PLAY

Create the illusion of bricks building themselves into a house by fixing a new toy brick between each recording.

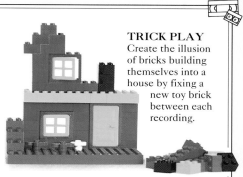

• STORYBOARD
Outline the composition and content of the shots, in advance.

STOP MOTION

Stop motion, the technique where by a series of very small movements are shot individually to make a "moving" sequence, works best with puppets or articulated dolls and toys. Any kind of movement can be simulated following this principle – only make a tiny movement between each recording. Shoot under an artificial light to avoid fluctuations in the **exposure**, and always use a tripod.

To walk – first move one leg   *Then move the body and one arm*   *Then move the other leg*

CARTOON ANIMATION

To shoot simple cartoon **animation**, use the animation or **interval timer** facilities. Set up a horizontal rostrum with the artwork placed on the floor or on a low table beneath the camcorder. Start your drawing and record your progress in short, successive, bursts. The playback will reveal the picture appearing as if by "magic" on a blank screen.

PICTURE •
A child can create a "magic" picture by drawing just one line between each recording.

Keep paper fixed in place

SKILL

13 CREATING TITLES

Definition: *Creating and making the most effective use of titles and captions*

MAIN TITLES, captions for individual sequences, and end credits are an effective way of shaping your video programs. Titles can be shot as part of a sequence either edited in-camera, insert edited later, or incorporated as part of a post-production process involving tape-to-tape editing. Program titles, captions, and credits can be superimposed over a background scene or simply shot as stand alone "artwork". Creating titles involves consideration for layout and technical factors, but most important is the aptness of the title design to the subject of your program.

OBJECTIVE: Maximizing the impact of your title design. *Rating* •••

CAPTION GENERATORS

Some camcorders incorporate a facility for generating text that can then be superimposed over your background scene while shooting. Choice of size, color, and position of text is limited, but designs should follow the general rules of layout. Pay attention to the relationship of the text color to that of your caption's background scene.

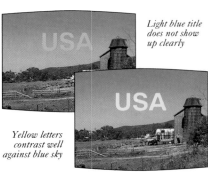

Light blue title does not show up clearly

Yellow letters contrast well against blue sky

Rocks in the foreground make the title cramped

Recompose to balance title against elements

LETTERING
When placing lettering over bright areas, such as sky, ensure that the color you use for your title or caption contrasts with the background scene sufficiently to make it clearly readable. Avoid pale colors on sky.

ADJUSTING COMPOSITION
You may need to adjust the composition of the text or background to achieve contrast and balance. The title (far left) appears awkward between the rocks and boat – achieve balance (left) by including only your title and the boat.

Use the names taken from the invitation

Take an opening shot of wedding cake

DIGITAL TITLES

Camcorders equipped with a digital memory store, allow titles or captions to be prepared in advance and then superimposed over the background scene while shooting. To provide an effective image for superimposition, titles need to be high contrast, such as black on white. Use a rostrum set up (see p.62) to shoot them, taking care they are straight and lighting is even.

COLORED LETTERING
Digital titling enables you to make full use of "found" titles and captions using a variety of methods, such as thick felt pens or hard copy from a word processor. Digital superimposers allow you to color letters for enhancement and ensure separation from the background.

BACKGROUND SCENES
When superimposing titles, ensure that the background scene is steady, since lettering is difficult to read where there is camera shake.

TYPOGRAPHY

For legibility, limit any text you want to use to six lines. Use simple, clean, bold letter styles. Avoid lettering with fine lines which can get lost against the background scene or the lines of the TV, and also elaborate lettering, which is too intricate to read quickly. Try not to mix type styles. Various sizes of the same style or upper and lower case can be used to emphasize differences. Avoid italics and being too ambitious with colored lettering.

Title is centered over some of the main image

Range title right to balance with main image

DOUBLE LINE TITLES
Double line titles can be centered or lined up to the left or right when they are balanced by elements in the background. Line spacing should be about half the height of the letters – this will ensure the words are not cramped.

SINGLE LINE TITLES
Generally, video titles look their best when they are simple, bold, and fairly large on the screen. A single line title is most apt centered on screen and set just below the middle line.

Set title just below center for balance

SKILL
13 FOUND TITLES

Bright shells make interesting lettering

When on location, you can improvise your titles and captions by using bold close-ups of signs, brochure covers, guide books, maps, etc. Titles can be scrawled in sand or snow, or lettering can be made up using shells, pebbles, flowers, or any other materials that you find around which are relevant to your topic. You can **animate** this method of lettering using the animation or the **time-lapse** modes – add just one item or letter between each shot.

• Flowers are ideal for nature subjects

LETTERING
When using materials such as flowers and shells to make words, ensure they are about the same size, so that your letters are neat and clearly legible.

ARTWORK TITLES

Artwork titles can be produced using a surprisingly vast range of materials. Make attractive titles by combining your lettering with illustrations or special drawings, such as those made by children. Alternatively, try simpler, but equally effective titles using rub-down lettering on colored card, or, perhaps, word processed titles photocopied onto sheets of colored paper. Children's printing sets, three dimensional letters, and hand drawn, cut-out letters, can also be used to produce impressive artwork titles.

STENCIL & RUB-DOWN LETTERS
Use stencil and rub-down lettering to give a neat look to titles. Draw a thin pencil line, (that can be erased later) as a layout guide. The ideal height for letters is 1cm (½in).

BLOCK LETTERS
Use three dimensional letters, similar to those found on children's building blocks, to make visually interesting titles – ideal for when making a video about children.

COMPUTER TITLES

Lettering and graphics on personal computer screens can be shot directly by the camcorder, although the image quality will be slightly degraded and may show **frame-beat** bars moving up the screen. The best image quality is obtained by linking the PC directly to an Edit VCR using an **encoder** to convert the signals. Use a **genlock** to combine the computer signals with live video signals from the camcorder and create superimposed titles. Paint and draw programs for home PCs enable you to create a variety of colorful titles and graphic forms.

• *Screen allows you to see layout immediately*

• *Use keyboard to input titles and captions.*

COMPUTER ANIMATION
The computer memory can be used to store title information that can then be **animated** by producing it on the screen in sequence.

EDITING TITLES

If you are planning to include captions and titles as part of in-camera editing, prepare them in advance or improvise on location with "found" titles. Insert editing them later gives you more control over their preparation and also their positioning. You will not be able to combine the titles and captions by superimposing them over a location scene, but you can still produce an effective substitute – such as using a holiday snap or even a postcard as the background scene for a title created with a caption generator or digital title superimposer. Main and end titles can be added to your videos in this way.

Original scene recorded on location

Digitally colored artwork title

Background scene and title combined

Title "reversed" out of the background scene

PRODUCTION MIXER
Adding titles at the tape-to-tape editing stage gives you maximum control of the position of the title scene in your video. By using a production mixer you can superimpose titles over chosen location shots and create a range of visual effects.

── TITLE LENGTHS ──
Do not hold main titles on the screen for more than approximately five seconds. For many short captions, three to four seconds is quite adequate. With blocks of text or lists of credits, calculate the time needed by reading the contents aloud to yourself at a normal reading speed – allow about a third as long again for slower readers. Titles held for too long will blunt the effect of your video.

SKILL

14

DAY 2

SOUND TRACKS

Definition: *The audio components contributing to a completed video*

USE SOUND to enhance your productions and provide important new elements in the overall effect. Add music or commentary, or both, to your edited program, using them as a replacement for the original location sound or blending the sources together into a single, mixed sound track. Sound can be used to complement or contrast with the pictures, providing your audience with the total experience that only the combined power of sound and vision allows.

OBJECTIVE: To create and record suitable sound tracks. *Rating* ●●●●

COMMENTARY

Commentary provides structure and continuity to a video, giving information that is not evident from the pictures alone

The Romans were great builders. They constructed temples, houses, and magnificent country buildings of carved marble. The Romans made great use of fired stone bricks too and developed concrete by mixing pozzolana, a tough strong volcanic form of material with rubble. Although they adopted many Greek architectural styles, they had their own great use of arches such as aqueducts. Romans also had skills in bringing water supplies to their cities along aqueducts. The Romans were good advanced civil engineers and built some of the first earliest arches known.

• **DRAFT SCRIPT**
Make a list of what you want to say, and where. Keep your language simple, informal, and free from jargon.

WRITING

Decide where commentary is needed and time each sequence. Write a draft script, allowing about three words per second as a reading speed. Keep your sentences short and avoid too many facts or figures as your audience will find these difficult to follow. Leave yourself plenty of breathing spaces – a commentary that runs through the whole video is tiring. Before actually recording, play back the sequences and read the script to check the fit.

INFORMATION
Always provide new or additional information in your commentary. Never simply describe what the pictures already show.

"The Romans were advanced civil engineers and built some of the earliest arches known"...

"Doric columns from a ruined temple built by the Greeks, in the country now known as Turkey"....

RECORDING

Ask someone to read the narrative for you while you control the recording. The reader should sound relaxed and speak clearly, keeping their level even. Record "to picture", watching the playback. Rehearse each section first while playing back, then put the VCR into audio dub mode in record/pause. Release the "pause" and give the reader a silent "cue" (such as a tap on the shoulder). If there is a mistake, go back to the start of the section and try again to ensure a good flow.

ACOUSTICS

Set up the recording in a quiet room that contains enough soft furnishings to ensure good acoustics. Draw the curtains closed to reduce reverberation from the windows and muffle any noises from outside.

• MIKE
Position an omni-directional mike 30-45cm (12-18in) from the speaker's mouth, or use a clip mike at chest level.

• TV MONITOR
Monitor the sequences as you record them to confirm that the commentary is correctly timed.

UNWANTED SOUND
Avoid recording machine sound from the VCR – turn down the TV, and keep the script paper still.

AUDIO DUB •
Audio dub enables easy replacement or re-recording of the separate sound tracks.

AUDIO DUBBING

Dubbing methods vary according to the format used. Camcorders and VCRs using formats with mono linear sound (VHS variants) enable you to replace the original recording by using the audio dub. Those machines with hi-fi tracks as well, allow the original hi-fi recording to be retained and the combined or separate elements to be heard during the playback. Formats that use only **FM** audio (8mm variants) cannot be treated in this way. However, the Hi8 camcorders with an additional **PCM** (Pulse Code Modulation) track, allow for dubbing on the PCM track. If two tracks are playing back together, the relative levels must be balanced. Stereo formats are only effective when played back through a stereo TV or hi-fi system.

Formats of 8mm with only FM tracks are difficult to use as the video and audio signals are recorded together and can only be separated as part of an editing or dubbing process

← *Video & audio FM*

← *Control tracks*
← *PCM sound*

SKILL

14

MUSIC & DUBBING

Music is the most potent element that you can introduce into your videos. It determines your audiences' response to the images they see

USING MUSIC

Most videos, be they of a wedding, a holiday, or a simple record of a day out, are enriched by the addition of a suitable music track. Music can either replace the location sound recording or be incorporated with it in a single mixed sound track. Your task is to find suitable music for the video – or create it.

Swan – slow and graceful

Horse – quick and lively

TEMPO
When using two pieces in consecutive sequences (as above), ensure their tempos clearly mark the changes in the sequences.

COPYRIGHT
Almost all commercially available recordings are covered by copyright, so using them to dub onto your videos will normally involve a form of consent. However, copyright holders are usually more concerned about clearances and due payments when public screenings or video sales are going to be involved.

CHOOSING MUSIC

Choose music by considering the tempo, style, mood, and length. Try out each potential piece by playing it back with the picture – adjusting the starting point in the music until you have the best fit. Opening and closing themes are the two most natural uses of music. These do not need to be long to be effective – fifteen to thirty seconds at the start and finish will help give any video a sense of shape. Try using short sections of the same theme to link different sequences.

EDITING TO MUSIC

Music can be used to great effect by editing the pictures to blend with the tempo of the music. Do this for an individual sequence or even a whole video, using insert edit facilities – either by using the controls of the Edit VCR or with the aid of an edit controller. Start by dubbing the music onto the linear sound track, play it back a few times until you are familiar with the tempos and changes, then start adding the pictures in sequence.

Snow-skier

Water-skier

MONTAGE
Use music to create a "montage" by linking a variety of shots all sharing a single theme.

MAKING MUSIC

Creating a special "score" is the ideal. It does not have to be very elaborate. Even if you have just basic skills on a single musical instrument you can create a simple accompaniment using either traditional themes or your own compositions. Whether your tape has been edited in-camera or by the tape-to-tape method, begin by timing the sequences that require music. Once a theme has been devised, try it out by playing it back with your tape.

ELECTRONIC MUSIC

The use of computer aided instruments such as electronic keyboards considerably extends the range of sounds you are able to produce.

SOUND MIXING

By using a simple sound mixer, it is possible to blend sound from several sources onto just a single track. If you have an edited in-camera video of, for instance, your holiday travels, you can mix the location sound together with a live commentary and music from an audio tape or CD while copying the video onto an Edit VCR. This will defeat the problem of dealing with tapes where your sound cannot be overdubbed. Mixing in this way can be further simplified by pre-recording the commentary onto an audio tape.

RESULT

While the picture is being copied directly from the camcorder to the Edit VCR, the original video sound track is fed into the sound mixer.

— ADDING MUSIC —

With the VHS formats (that have only linear sound tracks), recorded music can be dubbed directly onto the camcorder using the audio dub facility. However, music has to be recorded via the mike (by placing it just a few inches in front of a hi-fi loudspeaker) and the **ALC** system tends to level out the peaks and troughs in the volume of the music. Fading the music in or out is controlled from the "source" machine. The VHS VCRs with audio dub enable you to add music from any line source (tape, CD, or record) by plugging the source directly into the Edit VCR. Dubbing music onto VHS format tapes which have both a linear and hi-fi track, simply involves replacing the linear track – the original sound is still capable of being reproduced from the hi-fi track. However, this method of adding music may require more than one "take" of the dub to get the balance of the levels correct between the tracks. Music can only be dubbed onto 8mm formats that have a separate **PCM** track.

SIMPLE STUDIO

Once you are ready to record your music, create a simple studio – similar to that used for commentary recording. Use a dynamic mike for the instrument, or the line output from an electronic keyboard. Play back the video sequences as the recording is made. Ideally, record onto audio tape or cassette.

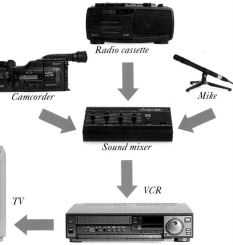

Radio cassette

Camcorder

Mike

Sound mixer

VCR

TV

AFTER THE WEEKEND

Exercising the skills you have learned to tackle popular subjects

WITH THE WEEKEND OF LEARNING behind you, sharpen up on the basic techniques that have been introduced. Try exercises to build your skills in hand-held shooting and moving with the camera by recording everyday events around your home. To improve your panning and tilting skills with moving subjects, record shots of a friend pacing from one position to another and walking up and down a flight of stairs, pausing at each turn. To explore the potential of composition, take a series of different static shots of the same

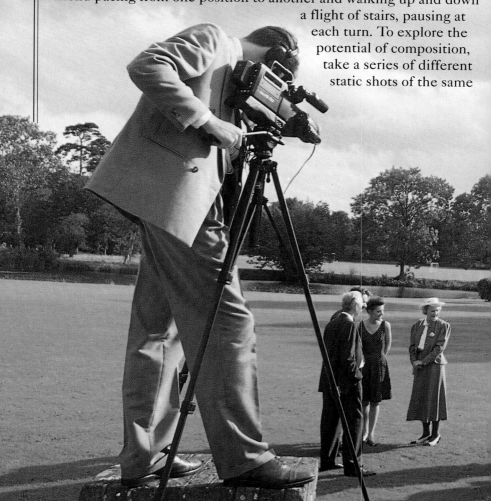

subject – a church or a single building in your own street, for instance. Test your understanding of screen direction and time by constructing a sequence of shots (using in-camera editing) showing someone going through the routine of setting off for work or school. Concentrate on getting eyelines correct by shooting a sequence of your friends or family having a meal around a table, and use a variety of camera positions and shot sizes. When the techniques that once seemed daunting become almost instinctive, then it is time for you to put your level of aptitude and talent to the test. The best way to do this is by setting yourself a project based on a theme that interests both you and your audience, and, most importantly, provides you with advanced and more elaborate challenges.

KEY MOMENTS
Weddings are one of the most important subjects for video-makers since your recording will be valued for many years. The challenge is to capture the key moments that cannot be repeated and produce a creative record of the event.

WEDDINGS

One of the most popular uses of a camcorder is "The Wedding"

SHOOTING A WEDDING offers you an ideal opportunity to put all the skills you have covered into practice. The key to a successful wedding video lies in careful planning and preparation. Your main concern is moving between locations in time – visit each place and time how long the trips take. This will help you plan sequences and simultaneously enable you to check lighting levels, direction of sunlight, power supplies, and acoustics (clap hands once and listen for the reverberation). Get permission to shoot, and take test shots on your initial visits. If there is a rehearsal for the church ceremony, attend it. Take test shots from a variety of positions and check the sound levels too. If there is no rehearsal, arrange for someone to go through the details of all the events (ceremony, reception, etc) in advance.

BEFORE THE DAY

Making all the necessary arrangements to help organize your shots and sequences in preparation for the "big day"

PLANNING

Make a list of all the topics and events you wish to cover, thinking about how you will shoot each one and then make them into coherent sequences. This is specially important if you are planning to edit in-camera. On your initial visit to the church, locate electrical outlets for lights, assess where a battery light may be useful, where to use auxiliary mikes, and when to use your tripod. Think about which filters will help to enhance each image. Ensure you know what is going to happen, and when.

SHOT LISTS
Write a shot list to help pre-visualize each sequence so that you are prepared for each event. The arrival of the bride at the church might be covered like this (right).

- Establishing shot of the church
- LS groom and best man at doorway
- CU of the groom smiling
- CU best man looking at watch
- I-telephoto shot of church clock
- LS groom and best man turning and walking into the church
- LS of the bridal car arriving at church
- Cut-away of all the onlookers
- Low angle (with filter) shot of the bride getting out of car, exits frame
- Pan with bride and father as they walk to the church doorway
- LS of the groom and best man as they turn and walk into church
- LS of the bridal car arriving
- Cut-away onlookers
- Low angle (with filter), bride just getting out of car, then exits frame

STORYBOARD

For key sequences, such as the ceremony itself, it will help greatly if you produce a storyboard. This consists of a series of sketches, visualizing each of the arranged shots, and will enable you to organize your shot positions and plan the way you will move between each of them.

1 Long shot of bride arriving at the church and getting out of the car. Pan right as she walks towards the church door

2 Long shot (taken from the inside of the church door) to show the bride and father entering the church

3 Wide shot (to include the priest and the congregation) as bride and father start walking up the aisle

4 Cut-in shot of the bride and groom exchanging their rings. Include the priest and some members of the congregation

5 Close-up shot of the bride receiving her wedding ring. Show detail of the groom holding the bride's hands

OPENING SHOTS

Plan the content of the opening shots and sequences in preparation for the wedding day. These could range from some simple pictures of the wedding invitation (use the **macro** facility p.23 and p.86) or a picturesque shot of the church with a superimposed caption (use the camcorder's caption facility pp.66-67), to more elaborate openings, such as taking still photographs from the family albums of the bride and groom (using rostrum skills p.62).

ROSTRUM SHOTS

Set the scene for the wedding day by taking a sequence of rostrum shots of the invitations and photos of the bride and groom as children.

Groom as a child

Bride as a child

Wedding cake

ON THE DAY

*These are the sequences of the important events for the
video-maker to follow through the wedding*

PREPARATIONS

An initial sequence could cover some
of the preparations on the day. This
might include decorations being set
out at the reception, the cake being
delivered, flowers being arranged in
the church, and the bride finalizing
her preparations at home. Cutting
between these events will create an
air of expectation. The shots could
be linked by a series of **whip pans**
to heighten the sense of excitement.

*Arrival of the
cake at the
reception*

*Flowers are
arranged for
the wedding*

*Bride prepares
herself for the
ceremony*

DEPARTING FROM HOME

Get a good shot of the bride and her escort
departing from home. Compose a shot of
the doorway and **zoom** out and pan as the
couple walk from the door to the car.

*Father and
bride leaving
the house*

*Low angle shot
of bride in the
bridal car*

THE BRIDAL CAR

Once the couple are in the car, move ahead
of the vehicle to a long shot position and get
a low angle showing the car driving off.

ARRIVAL

Ensure that you have plenty of time to
reach the church before the arrival of
the bride by arranging each of the pre-
ceremony sequences in advance. Also,
make sure you are in an advantageous
position to cover the guests arriving at
the church. Start with a well composed
static view so that all the guests simply
walk into shot and then pan with each
set of new guests as they arrive – then
cut back to your front position ready for
the next arrival. This is a good way to
avoid a series of **jump cuts**.

GUESTS' ARRIVAL

Choose a camera position
about 2m (6ft) off the path-
way leading to the door of
the church. As the guests
arrive, follow each of them
with a pan shot until they
have entered the building.

THE CEREMONY

The camera position determines how much of the ceremony you can cover. The ideal arrangement should give you a frontal or three-quarter view of the couple with the congregation in the background. The minister is then largely "back" to the camera so you will need to ensure that he does not mask your view of the couple at vital moments. Use a tripod and make sure you are far enough back to get a full length shot of the couple in the **wide angle** position. If you are restricted to this single station, you can still use it to take telephoto shots of the families and friends among the congregation. These **cut-aways** can then be used to bridge the sections of the service that you have not recorded.

Stay wide for the start of the ceremony

Tighten to a mid-shot to include guests

Tighten to a close shot of the bride holding the ring

Widen after the blessing to include family

SOUND

Use an auxiliary **cardiod** mike on a stand or taped to a lectern, as close as possible to the couple and priest – do not worry if it is in shot, but do ensure the cable is firmly taped down.

Mother fixing the bride for a photograph

The guests bid the couple a fond farewell

The bride getting into the bridal car

OUTSIDE CHURCH

To cover the couple as they emerge from the church, set up some distance from the church's doorway. Shoot a **reverse angle** of on-lookers as a **cut-away**, by swinging around, returning to a tighter view of the families and friends to end. For amusing, informal shots, try to get cut-aways of children involved in mischievous activity, and guests preparing for the professional photos. Use a tripod and the telephoto facility to cover the guests, family, and newly-weds from the single position.

DEPARTING

The newly-weds getting into the car and driving off is a key moment. Set up near the bridal car so that you start with the couple approaching you, then pan to follow them.

AT THE RECEPTION

Choosing and recording the key activities at the wedding reception – speeches, dancing, and eating

RECEIVING GUESTS

At receptions, even where receiving the guests is quite informal, you need to establish a camera position that will enable you to include both the bride and groom, the receiving family, and the guests as they enter the reception area. Set up the tripod, parallel to, and a couple of meters (6ft) away, from the receiving line. Start panning the guests as they arrive, bring the bride and groom into the shot as they greet them. This position provides the best angle from which to shoot all of the guests as and when they arrive.

The guests arriving at the reception

The guests being introduced to the couple

CAMERA HEIGHT
Set the camera slightly below the eyelevel of the guests to avoid faces being masked by hats or veils. Use a battery light on the camcorder.

SPEECHES

Wide angle to show family at head table

Zoom in as first speech commences

Tighten to close-up for highlights of the speech

Pan around to show the guests listening

When shooting the speeches, sound will naturally be a key consideration. Set up an auxiliary **cardiod** mike on a stand in front of the speakers. Even if the mike is in shot, it will give a good recording level and help to avoid the echo found in large rooms. The mike lead needs to be long enough for you to be far enough away to get a wide view of the proceedings. Set up on the tripod at about 45 degrees to the speakers. Get copies of speeches, or some knowledge of their content, in advance. If you are editing in-camera, incorporate all of your shot sizes and camera moves in a single sequence. Make these changes in a planned manner.

LIGHTING
Use video lights set well back against walls to provide a wide coverage. Set lights high on stands at 45 degrees to the speakers and weight the bases for safety. In smaller rooms, bounce the light off white ceilings.

CAKE CUTTING

Prepare for the cutting of the cake by getting composed shots of it before the actual "cutting". Use an on-board battery light for full illumination and the **macro** mode for detail shots of the cake's decorations. Adopt an elevated angle for the cutting ceremony.

A high angle for the cutting of the cake

Tilt up for close-up of the couple

ILLUMINATION
During the cake cutting, use a backlight on the newly-weds to add a romantic effect.

Shots of guests chatting informally

Candid moments between guests

INFORMAL SHOTS
Build a sequence by covering the wedding guests socializing informally together.

CANDID SHOTS

The coverage of the reception should include some shots of the individual guests. Use the telephoto lens to get candid close-ups of people. Find an initial camera position that gives you a clear view and use this for shots of different sizes and also for shooting in various directions. Wait for just the right moments to capture guests at their most expressive. This enables you to build up sequences of personal vignettes. Use the manual focus to target individuals within groups.

CREATIVITY
Dance sequences provide an opportunity for creative camera work that will contrast with the more formal elements of the event. Try varying shooting levels, shot sizes, and using filters.

TAPE-TO-TAPE EDITING

If you are planning to edit tape-to-tape, you will have greater freedom for shooting both different sized shots and **cut-aways** of guests – but you may then risk losing an important part of a speech. By making a separate "sound only" recording on audio tape, you will be able to dub this over some of the cut-aways. This will give you freedom to be selective with the content of the speeches. If you are shooting with two camcorders, use one to concentrate on long takes of the speakers while the other provides **reverse angle** shots of the guests – ensure you include both group shots and close-ups. These shots can then be insert edited later as part of the tape-to-tape editing process – providing you with the maximum creative choice.

HOLIDAYS & TRAVEL

A popular use of the video camera is to record one's traveling experiences

·

HOLIDAYS AND TRAVEL SUBJECTS provide some of the greatest opportunities for video-making. Although the possibilities of pre-planning individual sequences may be limited until arrival at the location, useful preparations can still be undertaken in advance. Before traveling to new countries or cities, read up about the area so that you have a clear idea of the kind of subjects that are likely to be of interest. Knowing what you would like to shoot will aid your mental preparation for the way you might need to shoot it. To a degree, the approach to the most common topics naturally follows a general pattern that reflects the subject matter itself.

BEACHES, TOWNS, & BUILDINGS

Making interesting and entertaining videos from recording scenes found at the beaches and towns you have visited

BEACH LIFE

Take time to compose views of your location and closer shots of people relaxing, but otherwise, concentrate on the more animated events. With children's activities, such as making sandcastles, build your shots up into natural sequences using **cut-ins** and **cut-aways** for structure. For shots of small children playing in shallow waves, walk into the water yourself so that you get views looking towards the beach as well as away from it.

ESTABLISHING VIEW
Set the scene with establishing location shots before concentrating on more lively activities.

SLOW MOVERS
Take static shots of slow movers. Keep the camera still and allow the subject to pass through the frame.

FAST MOVERS
Fast moving subjects, such as windsurfers and speed boats, demand telephoto lenses and stable tripod mounts.

Paraglider – static shot

Windsurfer – telephoto lens

TOWN LIFE

Before attempting a "portrait" of a small town or village, walk around it without taking any shots simply to get to know the place. Make notes on any worthwhile subjects and potential angles, and spend time sketching out a written shot list of the sequences. Plan the opening and closing images of the sequences and determine how you will link from one element to the next. Make the tempo of the sequence reflect the pace of the local life.

ACTIVITY SCENES
After an opening shot of the town, take some activity shots to build a montage (below).

Long shot of market scene

Mid-shot of customers

Close-up of a stall owner

Close-up detail of the building

Long shot of the building

ZOOMING
Zoom movements can be an effective embellishment for building shots. Try an introductory shot that zooms out from an interesting detail to reveal the striking appearance of the whole building.

BUILDINGS

Cathedrals, palaces, and castles are all natural topics for travel videos, but before you start shooting, take note of the light. Avoid shooting with the sun overhead at midday – it creates heavy shadows that will disguise details and the contrast of video accentuates this. The best time is when the sun slants across buildings – emphasizing their form. Build sequences from details.

— WIDE ANGLE —

Landscapes and townscapes, even from high vantage points, may be difficult to encompass using the standard **wide angle** position of your lens. To avoid the necessity of excessive hand-held panning, use an accessory wide angle converter. These increase the angle of vision by around 40 per cent.

TIME & WEATHER

*More ideas for other interesting subjects to record
while traveling or on holiday*

SUNSETS

Scenes shot at sunset or sunrise
are made more interesting if
you include a prominent visual
element in your shot as well as
strong coloring. Try to include a
striking foreground using figures,
buildings, or coastlines, even if only
seen in silhouette or, better still, some
form of activity – horse riders on the
sea shore, a sailboat passing, or birds
swirling in the sky. For reproduction
of colors, use manual white balance,
otherwise, your auto-system is likely
to "correct" the color balance by
"normalizing" the image. Also, where
possible, use manual **exposure**.

*When shooting sunsets,
keep to the same aperture
setting as the light fades*

SUNSET
Using manual iris settings, **zoom** in to a
portion of the sky to set a suitable aperture.

NIGHT SCENES

The camcorder's low light capabilities
can be exploited for night scenes and,
if necessary, "gain" controls (see p.19)
can be used to boost the **exposure** of
the image. City streets, particularly
with the extra lights around cafes and
shops, can produce adequate lighting
levels for recording after dark, but the
different **color temperatures** of the
various light sources, notably that of
sodium street lights, may result in
color casts within the image.

NIGHT EXTERIOR
Street and cafe lights can provide sufficient
illumination to enable shooting at night.

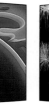

Neon lights

Firework display

ILLUMINATIONS
Illuminations and neon lights
(far left) can easily be treated
as abstract subjects – shoot
with star, diffusion, or color
filters. Alternatively, shoot
deliberately out of focus for
creative effect. Fairgrounds
and firework displays (left)
are best tackled by using the
manual iris settings.

SNOW

Bright sunlight and snow can create **exposure** problems. The overall brightness may be greater than the image sensor can deal with, resulting in **overexposed** images. One way to compensate is to use faster **shutter speeds**. Figures seen against snow or the skyline can appear **underexposed** – use the BLC or the manual exposure settings to boost the exposure. Battery capacity is reduced in cold conditions so make sure you have charged spares to hand.

Overexposed shot

*Use a ND filter to ensure correct **exposure***

• **CASING**
The damp/splash proof, heavy duty housing has large control buttons.

SPLASH PROOF HOUSING

Where there is a danger of spray, always use a camcorder with a splash proof housing to protect against damp and condensation.

CORRECT EXPOSURE

Exposure is a problem when shooting in very bright conditions with snow as the background (top left). Use an ND filter to reduce both the glare of the sun reflected off the snow and the depth of field, making the figure stand out clearly (top right).

NEUTRAL DENSITY FILTER

When using your camcorder in excessive or intense light, neutral density filters are essential as they help prevent **overexposure** problems.

TRAVELING WITH YOUR CAMCORDER

• Insurance – camcorders are a target for thieves. Insure your equipment fully and include third party cover in case of accidents involving other people.
• Customs – take all equipment purchase receipts with you for showing to your own customs on return journeys. Check with your travel agent whether you will need additional clearances for local officials.
• Inspection – X-ray machines or metal detectors cannot damage video tapes. However, you should be cautious with large electromagnetic machines at airports and elsewhere. If necessary, ask for your cases to be hand searched.
• Power – most battery chargers are automatically

or manually adjusted to accommodate the differing local voltages. Set these as appropriate, otherwise take a voltage adaptor.
• International Standards – there are three main color TV systems in operation worldwide – PAL, NTSC and SECAM. To play back tapes on TV while abroad, the local system must be the same as that of your camcorder and home system. Connections are then best made with A/V leads.

Remember – voltage adaptors may also be required for your plugs abroad

Transit – when traveling by air, carry your equipment as hand luggage in a strong protective case

PASTIMES

Using the camcorder to enhance your enjoyment of other interests

YOU CAN USE VIDEOS in conjunction with your other hobbies and special interests either by obtaining specialized video equipment or utilizing facilities you may already have. When shooting sports, decide on the kind of coverage you want – a record of the event, an impressionistic treatment that captures the spirit of the sport, personalities in the game, or an analysis of technique.

SHOOTING IN THE WET

A camcorder with a special splash proof housing will enable you to shoot a vast range of subjects that are far too risky to tackle with ordinary equipment. Water sports such as sailing, power-boating, and water skiing are just a few examples where these housings enable you to get spectacular angles without having to worry about the risk of spray. Snow subjects, such as skiing, tobagganing, even snowball fights and falling snow can be covered. Shooting in light rain and dusty areas also becomes possible.

UNDERWATER
Camcorders fitted with a waterproof housing open up the whole potential of underwater shooting.

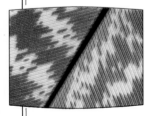

MACRO FOCUS
The **macro** facility allows extremely close-up focusing – enabling you to take shots of tiny subjects and also obtain shots of the smaller details of larger objects.

MICROSCOPE
The **macro** mode can be used with a microscope to obtain minutely, detailed close-up shots of subjects, such as this onion cell (right).

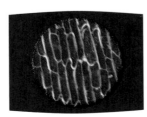

SEMI-SCIENTIFIC

The standard **macro** facility can be used for semi-scientific observation, such as bees pollinating a flower head. Depth of field is limited and lighting must be arranged carefully to avoid shadows, but with care, you can achieve striking pictures. Fast **shutter speeds** can also be used for slow-motion or freeze frame playback of natural phenomena. Using plenty of light and shutter speeds of one ten thousandth of a second, enables you to carry out effective, observational studies.

SPORTS

When covering any sport, the camera position is vital. For instance, to shoot a team sport, such as a soccer match, choose an elevated position that is close to the center line for wide shots that show the flow and strategy of the game. For action sequences, move nearer to the edge of the penalty area. Shoot everything from one side of the line, but if possible, move around to behind the goal to show corners and other goal mouth activity – remember to use manual focus behind the net.

RACES
Short races can be covered from a single position – near the finish line – using pan and **zoom**. Longer races allow changes of position.

TENNIS
Shoot court games from just one end of the court using a **wide angle**. Cover with gentle pans as the game moves from side to side.

MOUNTAIN BIKING
Course events can only be covered in part. Use telephoto and manual **zoom** outs for approach shots and long arc pans for passing.

SPORTS ANALYSIS

Video is an excellent tool for analyzing individual performance in sports such as golf, gymnastics, and diving. Use high speed shutters to record activities that can then be viewed and analyzed using freeze-frame and slow-motion playback with an appropriate VCR. Ideally, your instructor or coach should be involved in both the recording and the playback, but you can easily record your own performance. Set up the camcorder on a tripod and compose a shot that is wide enough to encompass the whole of the activity. Use manual focusing to pre-set the focus. Always use well lit conditions when using fast **shutter speeds**.

Back swing   *Forward swing*   *Follow through*

ANIMALS

Pets and wildlife provide worthwhile challenges

SUCCESS DEPENDS AS MUCH ON your knowledge of the animals as on video techniques. With your own pets, take advantage of your familiarity with their habits and routines to plan out sequences that will capture their personality and build them into a simple portrait video. If lighting is needed for shooting indoors, use reflected and diffused techniques for the gentlest effect and to avoid alarming the animals.

PETS

Dogs make good subjects since they can respond to commands. To create "point of view" tracking shots, carry your camcorder inside a plastic bag with a hole cut out for the lens. When filming cats, try to get down to their level. As they require more patience than dogs, be prepared for plenty of editing. Kittens and puppies are too unpredictable and energetic to try to follow – instead, pen them into a small area. Allow for retakes by rewinding the tape when shots go wrong.

• POSITION
Use toys to keep the kitten in one place and also to attract it into specific positions.

LEAPS & FALLS
Use a fast shutter and slow motion replay to analyze some movements – leaps and falls.

SKILLS & TRICKS
Create sequences with a pre-planned shot list to display your pet's skills and tricks.

BIRDLIFE

All birds found in the garden are best recorded when feeding. Either use a bird-table and establish their regular feeding times or use an unattended **interval timer** to check feeding time habits. There is no need to build an elaborate hide, simply shoot from indoors with the curtains closed and a chink left for the lens to poke out through. Use a tripod and a telephoto position. Vary shot sizes to create sequences and use a gun mike to record birdsong. By connecting your camcorder to the TV you can use a remote "control" for recording.

TELECONVERTER
A teleconverter is a real advantage for obtaining effective close-ups of bird life. However, do take care when operating the camcorder as even the slightest shake will be magnified.

AUTO-FOCUS
Bars and netting will confuse auto-focus – use manual control and stand well back from bars.

ZOO ANIMALS

Best coverage is had at feeding times when animal positions will be more predictable. A stable position allows you to pan when following the animals. The telephoto setting can be used for close-ups of sedentary animals, with an out of focus background – recreating the wild. Compose ahead of grazing herds and allow the animals to walk into shot. With very large animals, try to include people for scale and vary your shot heights to emphasize size.

WILDLIFE

In safari parks, all video shooting must be taken from inside a moving car. To prevent window reflections problems, use the manual focus. Motion makes the telephoto position difficult – shoot a three-quarter wide angle and lean forward or through the windscreen. For real safaris, use the teleconverter and tripod. Be prepared for long waits. Shoot early or late for advantageous lighting conditions. Keep equipment in the shade and cool, when not in use.

WATERING HOLE
Animals naturally gravitate to watering places – providing an ideal opportunity for filming.

AVOIDING PROBLEMS

Ensuring proper care and maintenance of your camcorder

CAMCORDERS REQUIRE PROPER CARE, regular maintenance, and occasional servicing. The complexities of the optics, electronics, and mechanical parts mean that expert attention is required when repairs are necessary, which can be expensive. Help avoid these costs by being aware of the functions of the camcorder and the type of problems that can arise during regular use. Follow the manufacturer's instructions and take basic precautions to help ensure trouble-free use.

TROUBLE SHOOTING

Problems can arise even when you use the camcorder correctly. However, most of these are due to malfunctions and are simple to deal with. Camcorders have auto cut-outs to help prevent problems. The power cuts-out if the tape transport reaches a part of the tape that is damaged and will also turn off if the machine is left in pause for more than a few minutes.

Battery compartment •

POWER PROBLEMS
If the camcorder will not function – check the battery is not flat, that there is tape in the machine, and the power is on.

Power Capsule •

• Battery

Mike •

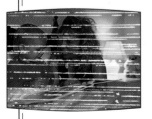

IMAGE
This type of image distortion is caused by dirty tape heads. Other possible causes for unclear images can be due to condensation on viewfinder or lens.

SOUND
If the quality of sound is poor or nonexistent, check that the external mike is switched on, the battery is not dead, and the heads are clean. A "howl" sound indicates it is operating too close to the TV – turn down the sound.

-PLAYBACK PROBLEMS-

Main playback problems are usually caused by the following situations:
Tape stops on fast rewind – check that the memory is not switched on.
Tape will not rewind – camcorder may be set to the record/camera function.
Cannot remove tape – battery is dead.

RECORDING
If camcorder will not record – ensure the tape's safety tab is in place. Power failing quickly can be due to excessive use of motors and operating in cold.

Crack-off safety tab •

C14P7118   MADE IN JAPAN

• Slide over safety tab

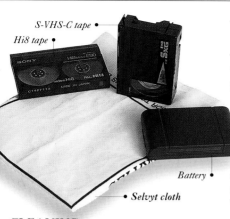

S-VHS-C tape •

Hi8 tape •

Battery •

• Selvyt cloth

EQUIPMENT CARE

To care for your camcorder, always try to avoid sand, dust and damp, shocks, heavy vibrations, magnetic fields, water, indirect heat (hot cars), and direct heat (from lamps). Never carry the camcorder by its lens, mike, or viewfinder – always carry by its handle. Do not force tapes in or out of the machine. Remember to remove battery and tape when not in use.

CLEANING
Clean the tape heads regularly. Use cotton swabs to clean the viewfinder screen. Use a selvyt cloth to clean the surfaces.

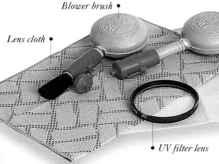

Blower brush •

Lens cloth •

• Lens cap

• UV filter lens

LENSES
Avoid strong lights on the lens. Keep the cap on when not in use. Protect with a UV filter when shooting. Use a blower brush for cleaning.

Battery •

Battery • charger

• Protective video cases with labels

Battery •

TAPES
Keep tapes away from strong sunlight. Store tapes, rewound and in cases, standing upright at room temperature in a dust-free environment.

BATTERIES
Use batteries of the specified voltage and never short circuit or use them for running other equipment. Store batteries in a cool place and discharge them fully before recharging.

GLOSSARY

Words in *italic* are glossary entries.

A

• **ALC** Automatic Level Control. A facility that is controlled by electronic circuitry to automatically balance the sound recording level.

• **Animate/Animation** Recording inanimate subjects in different positions for durations of a few frames to create the illusion of movement.

B

• **Blue gel** Blue colored gelatine used on lights to bring tungsten light to the *color temperature* of daylight.

C

• **Cardiod** The heart shaped polar response pattern of a specific mike – sometimes called uni-directional mike.

• **Clean edits** Edits not visible when played back – accomplished when *synchronizing pulses* are all uniform.

• **Color temperature** The measure of the precise make-up of color provided by a specific light source. It effects the ability of the camcorder to reproduce colors accurately.

• **Crabbing** Method used for moving sideways in an arc around a subject when hand-holding the camcorder.

• **Cut-away** A shot used to cover the passing of time. It must be related to the main action, but subsidiary to it.

• **Cut-in** A shot used as a transition which forms a detailed close-up of a subject previously seen in a wide view.

E

• **Encoder** An electronic device used to convert the signals from a computer for recording onto video.

• **Exposure** The amount of light passing through the lens and reaching the image sensor. The *exposure* is controlled by the iris.

F

• **Filters** Disc of colored glass or gelatine, which fits over the lens and modifies the light passing through it. Different filters have various effects.

• **Fluid Head** An oil-dampened mechanism used on tripods and monopods to help ensure the smooth operation of camera movements.

• **Flying Erase Head** An erase head which is incorporated in the rotating drum of the recorder. It ensures the production of perfect electronic edits.

• **FM** Frequency Modulation. The recording system employed to produce hi-fi sound on camcorders and VCRs.

• **Focal Length** This is the distance between the optical center of the *zoom* lens and the image sensor. The longer the focal length, the greater the magnification involved, the shorter the focal length, the wider the angle of view.

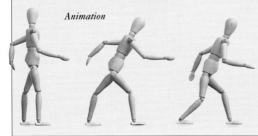

Animation

• **Frame-beat** Visible bar across the screen that becomes evident when videoing from VDU screen. Caused by *unsynchronized pulses*.

• **F-stops** Number that indicates the relative aperture of a lens at different diaphragm settings. The smaller the aperture, the higher the f-stop or number.

G

• **Genlock** An electronic locking mechanism that enables the combining of images from two different sources.

H

• **Helical scan** The recording system in which a rotating drum records a long, diagonal series of tracks from the video heads on a laterally moving tape.

• **Hold** The static composition that forms the beginning or end of a camera movement or a *zoom* shot.

I

• **Interval timer** A camcorder facility which records shots at timed intervals. Enables the production of *time-lapse* effects and crude *animation*.

J

• **Jump Cuts** A cut in which a portion of the action is noticeably omitted.

M

• **Macro** Facility for shooting extreme close-ups to give a magnified image.

O

• **Overexposure** Allowing too much light to reach the image sensor. It results in an excessively light image.

P

• **PCM** Pulse Modulation Code. A system of recording digital audio signals onto a separate track.

Q

• **Quick-release** Mechanism that forms part of the tripod's head enabling the camcorder to be swiftly mounted and dismounted.

R

• **Reverse angle** A shot taken from a camera position opposite to that used in the previous shot – reinforcing the relationship between the subjects.

S

• **Shutter speed** Electronic control of the duration of scanning – normally one sixtieth of a second. Fast shutter speeds result in reduced apertures.

• **Synchronizing pulses** Impulse on each video frame to control the speed and scan of the tape during playback.

• **S. Terminal** A connector used to ensure picture quality by keeping both the chroma and luminance signals separate from one another.

T

• **Time-lapse** Technique of exposing in short bursts of recording at regular intervals so that the event being shot appears speeded up when played back.

• **Time code** A frame-by-frame time reference incorporated within the recording to enable precise editing. It only becomes visible when played back for use with an edit controller.

U

• **Underexposure** Insufficient light reaching the image sensor which results in a dark image.

W

• **Wide angle** Angle produced with a short *focal length* lens which provides the widest angle of view and the greatest depth of field.

• **Whip pans** Rapid pan movement at the end of a shot that results in a blur. Short sections (less than ½ a second) can be edited to link related subjects.

Z

• **Zoom** Is the change of image size achieved with the zoom lens as the focal length is altered. The movement may be motorized or manual.

Quick-release

INDEX

GETTING IN TOUCH

Institute of Amateur Cinematographers,
24c West Street, Epsom, Surrey,
KT18 7RJ England

The Institute of Videography,
446 Burnley Road, Accrington, Lancs,
BB5 6JU England

You may be able to find clubs or organizations in your local Yellow Pages

ACKNOWLEDGMENTS

Roland Lewis and Dorling Kindersley would like to thank the following for
their help in the production of this book:

Damian Rodgett and Debbie Charles, Tracey, Georgina and Max Powls, and
Barbara Munns for modelling. Additional modelling by Georgia Bulmer,
R. C. Saunders, Rochester Motor Club, Mr and Mrs G. Wheeler,
the DK Weekend Team, Julia Heyes, and Brenda Wooding.

Dawn Lane for styling and hair and make-up, and Emma Kotch for hair
and make-up. Dave Leggate for the house location.
Buckmore Park Cart Circuit, Hever Castle Limited, the Cutty Sark and
the Maritime Trust, J Sainsbury plc for the trolley, and the Royal Horticultural Society.
Bridesmaids dresses supplied by Superior Tat, Cheam. Wedding cake supplied by
Heriot Catering, Enfield. Wedding tableware supplied by Regal Hire.

Equipment loaned by: Aico International Group, Audio Technica Ltd,
Canon (UK) Ltd, The Cygnet Photographic Co, C.Z. Scientific Instruments Ltd,
Fox Talbot, Hama PVAC Ltd, Hanimex (UK) Ltd, Introphoto Ltd, JVC (UK) Ltd,
Keith Johnson + Pelling Ltd, Johnsons of Hendon Ltd, Kodak Ltd, Lamba plc,
Panasonic (UK) Ltd, Sennheiser (UK) Ltd, and Chris Baker of Sony (UK) Ltd.

Sam Grimmer for design assistance and Tony Mudd for editorial assistance.
Hilary Bird for the index. Janos Marffy for color illustrations and line drawings.

Picture Credits: Roland Lewis p.14 (tl), p.25 (bl), p.63 (c, cr, bl, & bc), p.66,
p.67 (br & bl), Musée d'Orsay, Paris – Le Déjeuner, Claude Monet p.62 (b),
Commodore p.69 (tr), Paul Springett p.82 (cr), p.84 (cl), p.86 (tr),
E. V. Green p.82 (bl), Sam Grimmer p.82 (br), p.87 (cr)
Agency Credits: NHPA; M. Wendler p.42 (bl), p.89 (cl); Robert Harding.
Picture Library; p.43 (tl). ZEFA; p.84 (bc), Spichtinger p.87 (cl), Stockmarket p.88 (br),
Wegler p.88 (bl), Weir; p.69 (cla & bl), p.69 (bc). The Image Bank; R. Phillips p.83 (tr),
ACE; B. Simmons p.83 (cla, ca & cra), Mauritus Bildagentur p.85 (tc & cra),
P. Craven p.87 (tr). Bruce Coleman Ltd; H.Reinhard p.89 (tr).
t: top, b: bottom, l: left, r: right, c: centre, a: above

A note from the Author: I would like to express my thanks to the editorial and
art team at Dorling Kindersley, particularly to Alison Donovan for her creativity, hard
work, and patience. My very special thanks to FZT for all
her support, help, and encouragement over many months.